The Seashore World

The land is dearer for the sea,

The ocean for the shore.

Lucy Larcom, "On the Beach"

Vertical cliffs are resting places for pelicans and cormorants, which fly from their high perches to fish in the nearby sea. The pelicans dive from the air and the cormorants swim underwater to catch fish.

The
Seashore
World

David F. Costello

ILLUSTRATED BY THE AUTHOR

THOMAS Y. CROWELL, PUBLISHERS
Established 1834
New York

The photographs on pages 42 and 185 are reproduced courtesy of the Oregon Fish and Wildlife Commission; the photograph on page 76, courtesy of the News Bureau, Florida Department of Commerce. All other photographs are by the author.

FIRST EDITION

Designed by Lydia Link

U.S. Library of Congress Cataloging in Publication Data

Costello, David Francis, birth date
    The seashore world.
    Bibliography: p.
    Includes index.
    1. Seashore biology—United States.   I. Title.
QH95.7.C67 1980   574.973   79-7641
ISBN-0-690-01235-7

80 81 82 83 84 10 9 8 7 6 5 4 3 2 1

Contents

Acknowledgments

Although I have wandered up and down the seashores of America for many years, my personal experiences by themselves could never give a full grasp of the whole field of physical and biological phenomena of that narrow zone between land and sea. For many of the things I have seen and the places I have visited I am indebted to friends and others who are knowledgeable about seashores near their homes. I am also indebted to hundreds of authors who have written technical books and scientific articles about the oceans and their shores. Their publications have increased my knowledge of seashore life and my understanding of how much is yet to be learned about the mysteries at the edge of the sea.

This book is not a scientific treatise. It is a sampling of some of the many aspects of the seashore that create a sense of wonder in the minds of people who visit the edge of the continent. If it induces some readers to look more closely at the teeming life forms at the edge of the sea, or to seek information beyond what is provided here, this book will have served its purpose.

I owe a debt, for inspiration and knowledge obtained by correspondence, to many people who live along the Atlantic, Gulf, and Pacific coasts. My knowledge has also been increased by the efforts of many people who have developed seaside aquariums and marine research stations. I remember particularly how Dr. D. J. Crisp, University College of North Wales, Bangor, took a few of us on a visit to the Marine Science Laboratories at Menai Bridge in April 1962.

It was late at night, but the labs were lit up from top to bottom. The researchers there were so absorbed with their studies of sea animals that night or day seemed to have little meaning for them.

Grateful acknowledgment is made to those who have helped me during the writing of this book. Marie Spivey, Acting Chief, Library Branch, Technical Information Center, Waterways Experiment Station, Corps of Engineers, Vicksburg, Mississippi, made reports from the library available and helped me obtain retention copies of the "National Shoreline Study, Inventory Reports," for various coastal regions of the United States. R. Duncan Morrow, National Park Service, Washington, D.C., kept me informed on basic information about national seashores. Bob Kuhn, Oregon Department of Fish and Wildlife, Portland, aided in locating picture material of Pacific coast animals. Kelco Division of Merck and Company, San Diego, California, provided information on kelp harvesting and sea urchins.

In a more personal way, Dr. Joyce Baldwin Costello, San Diego City College, San Diego, California, took me on excursions to find and photograph certain sea creatures not in my photo files. Edward J. Dortignac, Bowie, Maryland, skippered his boat around Chesapeake Bay so I could appreciate its vastness and innumerable estuarine variations. Mr. and Mrs. P.F.W. Prater of Tallahassee, Florida, took my wife and me to secluded places on the Florida Gulf shore where I could collect shells and study sea oats on the dunes. Mrs. C. A. McAlister and Mrs. Charles F. Ziebarth, Oceanside, Oregon, fed the gulls so I could photograph them and led me to choice agate collecting areas on the Pacific shore. John K. Terres, who edited four of my books for the Living World Series, gave me courage and understanding when I was in a writer's slump and told me about places of interest on Long Island, New York, and the remarkable dune areas at Seashore Natural Area at Cape Henry in Virginia Beach. And not least of all is my debt to Hugh Rawson, editor for this book, for advice, encouragement, assistance, and great human understanding of the trials and tribulations of an author.

David F. Costello

Fort Collins, Colorado

Arch rocks exemplify the power of the sea. Waves cut through less resistant zones of large rocks or through cracks. Freezing and thawing also accelerates the rate of disintegration. Eventually the roof of the arch falls and two sea stacks or rock islands remain until the waves wear them down below tide level.

1

The Varied World of the Seashore

ONE WINTER DAY NEAR MAXWELL POINT west of Tillamook, Oregon, I met a man collecting stones beside the surf of the Pacific Ocean. He had a plastic pail filled with agates and jade. He had more agates than any other beachcomber I had ever met at that place. Why was he so successful?

"Variety," he said. "There is always variety along the coast. Even the same place is never the same from day to day and from year to year. The waves are forever shifting the sands and rocks of the seashore. You have to understand the waves and the seasons to know when there is a poor time and a good time for collecting."

His words epitomized the changeableness of the seashore and its revelation of new artifacts with the passage of time. If he had been a shell collector on Padre Island, off the coast of Texas, he undoubtedly would have been aware that the greatest collecting time comes after a hurricane has raised giant tides that bring mollusks up on the sands in almost unbelievable variety. And if he had been a fisherman in a shallow Georgia bay, he would have known that a good time for fishing is when the fiddler crabs are fiddling and the great white herons are wading in water above their knees. But these are only local variations of the magical coastline of our nation.

Our thousands of miles of national seashore are replete with variety—from little tide pools housing their minute worlds of algae, crabs, and fish to diversified ecosystems on sandy shores, in mighty estuaries, and on rocky cliffs moistened by morning fog and buffeted by boisterous winds from endless miles of ocean. We are drawn to the seashore to see and enjoy this natural diversity, and to escape the monotony of towns, cities, and cultivated fields. We wish to revel in a world of extraordinary variety and complexity. Doubly blessed is the person who enjoys the little diversities of the seashore and at the same time looks at the broad view of the endless miles of differing terrain born of conflict between the sea and the land. Only then does one truly appreciate the majesty and the enchantment of that narrow strip of world where the foundation of the earth rises from the ocean and becomes the edge of the continent.

The variety of the seashore includes more than its landscapes. It includes the many moods of the sea. It includes innumerable environments for living things. And it offers opportunities for adventure with wind and waves and sand and animal and plant life each time we visit the shore.

The beach that yesterday was scarred by footprints and sand castles now is washed clean by the restless sea. The tide that hours ago ridged the flotsam of the previous night in windrows is out, and the attenuated thunder of the pounding surf seems far away. The gulls that soared effortlessly in the wind now patrol for stranded crabs and barnacles on the wave-rippled strand. On sandy beaches, the waves speed up the incline pushing the sanderlings before them. Then, on twinkling feet, the bird carpet flows seaward with the receding wave. This is the sea in one of its gentler moods.

The sea has other temperaments. At the feet of rocky cliffs there is no murmuring of the surf. Instead, the breakers thunder against the rocks where starfish cling to their stony support until the tide relents and they can search for mussels in boulder-strewn tide pools. When the gentle winds blow in from the Gulf to the coast of Florida, the ghost crabs scuttle from grass clump to grass clump as fast as the eye can follow. Here is a land of gleaming white sand, tranquil lagoons, and narrow barrier islands built in former times by hurricane-driven waves and made picturesque by semitropical vegetation. Westward, along the Gulf shore, the coastline is a land of sandy beaches, coastal forests, and meandering streams. Endless marshes are populated with an extravaganza of bird, reptile, and other animal life.

It is estimated that the Atlantic, Pacific, and Arctic coastlines of the nation total 88,633 miles. In this incredibly long boundary of our part of the continent, it is inevitable that the variety of the seashore would derive from the endless permutations of geologic, climatic, and biological forces. Geological and shore dynamics have given us the glacier-scoured rocks and indented shorelines of Maine. Winds, storms, and tides have built the islands of Virginia and the Carolinas and the sand beaches of the Golden Isles of Georgia. Geological subsidence and uplift of the land have given us the Delta swamps of the Mississippi and the fringing desert along the Pacific Ocean and southward along the coast of Baja California. And not least of all, faulting and contortions of the earth's crust have produced the mountains and rocky prominences that rise from the sea from California to Washington. Interspersed with these are fjords, those cliff-bordered channels to the sea, lagoons, and estuaries both great and small.

The landward side of the coastal zone is a region of variable width. The edge is narrow at the base of rocky promontories and at the foot of mountains whose slopes plunge precipitously into the depths of the sea. In contrast, along the Atlantic and Gulf coasts, flat marshlands extend inland from the ocean boundary for many miles. Here, the interplay between land and sea does not stop at the water's edge. Salt spray penetrates inland and influences the saline content of the soil and the vegetation. The tides rise and fall twice daily in some places and once daily in others. They mix with the fresh water of estuaries and marshes and, in some instances, extend far upstream beyond salt water. The Pacific tide, for example, occurs in the Columbia River even beyond Portland, Oregon, more than 100 miles from the sea.

The seaward boundary of the South Atlantic coastal zone begins near the water's edge and extends many miles from shore. From the land of sand beaches, dunes, marshes, and wooded shores, the area of shallow water reaches to the edge of the continental shelf. This area of shallow water is the littoral zone which reaches from the high-tide mark to a depth of approximately 600 feet and contains some of the richest fishing areas of the world. Here, and on the landward edge of the ocean, are representatives of most of the animal kingdom and a multitudinous display of plant life. These mutually combine into ecosystems. Together, under the influence of differing physical, chemical, climatic, and topographic forces, they produce a seashore world with a thousand faces.

Seashore landscapes are the consummation of nature's fabrication of earth materials through millennia and even eons of time. Recognition of the various kinds of coastlines and the processes that formed them adds immeasurably to our appreciation of their uniqueness in nature and their artistic appeal when we visit the seashore. If we look thoughtfully at the topography of seashores, it becomes apparent that their landscapes are products of the forces that shape the land and of the sculpturing action of the sea.

Many of our coastal seashore landscapes owe their forms principally to the land itself. Rocky cliffs, drowned river valleys, moraines left by glaciers, and deltas formed by river deposits are made primarily of the substances of the land and not of the sea. Marine processes have left these essentially unchanged. On the other hand, barrier islands, lagoons, and inlets, such as those in the vicinity of Hatteras Island and the Outer Banks of the Atlantic coast, are largely the work of the sea. The sea stacks or rock islands, left as erosional remnants of promontories along the Pacific coast, also are the work of the tides and waves.

The sandy lands of the seashore appeal to countless people because of their accessibility and the opportunity they provide for intimacy with the sea. They have so much variety and are so volatile under the influence of the waves and the wind that they are forever new. If one considers their origin, their changing beauty, and their environments for living creatures, they provide endless opportunity for enjoyment of physical and biological nature.

Sandy lands owe their existence to the travels of individual sand grains. Freezing, erosion by wind and water, and gravity pry apart the rocks of mountains. These particles of earth material are carried by rivers to the sea, from which they originally came. The action of water and waves grinds them into smaller and smaller fragments and deposits them in bars, sandspits, and berms (narrow sandbanks) along the seashore. Sands deposited above the waterline by the swash and backwash of the waves are picked up, sorted by the wind, and fashioned into dunes or built into barrier islands, separating the mainland from the sea with shallow lagoons that eventually become coastal marshes.

Sand moves beneath the surface of the sea as well as on land. Ocean currents drift sand into bars that gradually rise above the surface to become sandspits, or long narrow ridges, that enclose bays or enlarge islands with accumulated deposits. Netarts and Nehalem bays in Oregon are good examples of bays that are protected from the direct action of ocean waves by sandspits. An outstanding sandspit exists at Provincetown Harbor at the tip of Cape Cod, where the sand has drifted from the Highlands of Truro.

Many other sandy coasts owe their presence to the movement of material along the shore. When wind and waves bring sand landward at an angle with the beach, the sand returns straight out with the backwash. The result is that it drifts longitudinally along the shore. It also moves seasonally with changes in the wave action of the sea. The summer berm, built by gentle waves, is removed by storms or is blown inland to form dunes. The winter berm is formed by waves that move sand and debris high on the beach.

Extensive sand beaches occur along the coast south of Cape Cod. The sea islands of New Jersey, Delaware, and Maryland are long barriers separated from the mainland by shallow lagoons. The great recreational areas of the Carolinas include the sandy beaches of capes Hatteras, Lookout, and Fear in North Carolina and Cape Romain in South Carolina. Sandy islands border the mainland of South Carolina and Georgia, while, along the east coast of Florida, the barrier islands are narrow and extend for dozens of miles without inlets from the sea.

Sandy barrier islands in the Gulf of Mexico occur from Florida to Texas. Some of these have the most attractive beaches found along any of our shores. Sanibel Island, west of Florida, is a shell collector's paradise. The gleaming white beaches of Santa Rosa Island near Pensacola Bay are matched nowhere for their nearly pure quartz sand. Padre Island, the largest of the Texas sand barriers, is notable because of its continuous stretches of fine sand, its bird life, and its legends of pirates and buried treasure.

The Pacific coast, from Baja California to the San Juan Islands in Washington, presents a panorama of mountains and lowlands bordering the sea with spits and broad beaches, rocky headlands, spit-enclosed bays, sand dunes, and long stretches of straight beaches. The majesty and extreme variety of this 2,000-mile stretch of Pacific coastline make it one of the most attractive recreational areas on the continent.

An animal world that we never see in its entirety exists on and in the sand, itself a utopia for burrowing creatures. The lower beach especially is crowded with bivalves, including cockles and various clams, an endless assortment of segmented worms, moon snails, sand shrimps, and crabs, and a multitude of algae, diatoms, and microscopic protozoans. These animals live in holes, tubes, burrows, and even between the sand grains themselves, where they are protected from sunlight and the dessicating winds. Later, we will examine some of these living forms in detail and see how they fit into food chains and food webs of the seashore.

Life on the upper beach where the sands are drier is sparse and generally consists of air-breathing animals that live in or just above the beach vegetation. Ghost crabs, tiger beetles, mole crickets, grasshoppers, ants, foxes, mice, and toads live here or venture down from the beach's landward edge.

Visitors to sandy shores know that when topography, wind direction, and the climatic pattern are right, dune lands are formed by sand that has been blown inland from berms along the beach. These great mounds of sand roam the shore like waves roam the ocean. They cover marshes, destroy forests, and then migrate, uncovering the stark skeletons of long buried trees. The dune environment is a harsh one where only specialized plants can grow. The vegetation itself greatly influences dune formation by disrupting air flow, thereby initiating small dunes which grow into larger ones. Far back from the sea, the dunes become stabilized when grasses, shrubs, and forests still the movement of the drifting sands. Much more will be said about these dynamic and fascinating landscapes in Chapter 6.

In contrast with the shifting sands of the dunes and beaches, the sea cliffs and rocky shores seem immovable, even though they are exposed to the continued violence of the sea. The sea eventually wins, for time is on its side. But the process is slow. On the Maine coast, for example, the granite ledges, scraped clean by the last glaciation, still extend into the sea as promontories and islands. On the Pacific coast, softer rocks, such as shales, sandstones, and lava, erode while the harder rocks persist as headlands or sea stacks, rising out of the water beyond the lower tide line.

In this difficult and scenic terrain, where some cliffs challenge even the most expert mountaineers, innumerable habitats permit the existence of varied plant and animal life. Boulders at the base of cliffs are pounded by the surf on their seaward side and protected from harsh wave action on their landward side. On their smooth faces, limpets, chitons, barnacles, and rockweeds endure the crushing force of the waves while sea anemones, hydroids, mollusks, and crabs flourish in the protection of clefts and crannies between the rocks. Grottoes and trenches that remain moist between the tides support a rich fauna of crabs and other creeping and crawling creatures. Above the slippery, algae-covered rocks the barnacles and mussels reach their upper limit, and land animals such as insects, snails, isopods, and spiders occupy the rocky surface.

As the sea cliff reaches higher and higher, ledges, gullies, crevices, and boulders with many faces offer footholds for lichens, mosses, ferns, and shrubs. On cliffs and sloping surfaces of promontories where plants and soil are established, seabirds create homes and raise their young. The nearness of the water ameliorates the climate and the sea provides their food.

Among the notable sea cliffs and rocky headlands are California's cliffs of Big Sur, where the sweep of the landscape extends upward from the sea to the slopes and tops of coastal mountains. Other scenic cliffs and tide rocks add variety to the seashore along the Mendocino coast, the San Mateo coast, and at Point Lobos, Point Reyes, and Tomales Point. Near Oceanside, Oregon, the Three Arch Rocks, the numerous sea stacks, the rock-studded headlands, and the barren cliffs attest to the ancient and continuing battle between land and sea. Farther north along the Pacific, the area of Olympic National Park is a rockbound coast that is relatively undisturbed by man.

In contrast with the abruptness of cliffs and promontories, some of the most distinctive transition areas between land and sea occur in estuaries, bays, and seacoast marshes. These watery habitats provide great diversity of living conditions for plants and animals. This is particularly true of estuaries, which are river mouths where tidal fluctuations contribute to the mixture of fresh water with salt water from the sea. Here, physical, chemical, and biological systems are interlocked in an environment that supports fascinating life cycles and food webs that are important to nature and man alike. These are the places to see and learn of alewives, salmon, oysters, crabs, shrimp, rails, sandpipers, and sea nurseries for fish of rivers and seas. These most vulnerable of all the seashore landscapes also are scenes of man's profligate exploitation and disregard of their value as buffer zones against the storms and tides, as pollution cleansing agents, and havens of unceasing color and beauty throughout the year.

Many of the world's estuaries now are seaports and centers of industry. Thousands of acres of marshes have been drained for use as agricultural lands or for urban expansion. Some, such as the nearly inaccessible coastal areas of Louisiana, are still essentially pristine. These offer livelihood for fishermen, protect migratory birds, and serve as buffers when hurricanes and storms stir up the

waves and tides to mighty proportions. Most estuaries and their adjoining marshlands provide recreation for the adventuresome and for
nature enthusiasts. Even more important, they provide unlimited
study material for scientists who seek to understand the marsh-estu-
ary system and how to manage it for the welfare of mankind.

There is great variety among the estuaries along the Atlantic seacoast. New York Harbor is a water-filled fjord cut by ancient glaciers. Delaware Bay and Chesapeake Bay, on the other hand, are
drowned river valleys with wide entrances to the sea and consequently are subject to strong tidal currents. Within these large estuaries are many embayments with narrow inlets and freshwater flow
from large and small rivers. Along with the estuaries, the great salt
meadows along the Atlantic and Gulf coasts have developed out of
shallow bays, on submerged sand flats, and in lagoons that have filled
with stream sediments and sand from tidal currents. Salt meadows
are rich in living organisms with intricate relationships and complicated food webs. They are temporary homes for many ocean dwellers that swim into their freshwater brooks and ponds to spawn and
then return to the sea.

The estuaries of Georgia and many of those along the Gulf coast
connect with the sea through bodies of water (sounds) between barrier islands. The Gulf coast, from Cape Sable to the Mississippi
delta, is a 700-mile mixture of alternating white-sand beaches,
grassy marshes, and tree-studded swamps. The largest estuaries are
Charlotte Harbor and Tampa Bay in Florida and Mobile Bay in
Alabama. The Louisiana shore has many sounds, bays, bayous, and
marshes. The marshes are surmounted in some localities by chênières, ancient beaches not covered by the postglacial rise in sea level.
The coastal bays with their mixture of brackish and marine waters
make Louisiana one of the leading producers of shrimp and lobster.

The Texas estuaries, Galveston, Matagorda, San Antonio, Aransas, and Corpus Christi, support a great recreation and fisheries industry. Here, brown shrimp, white shrimp, oysters, large-scale
menhaden, striped mullet, and other species find nursery space for
their young. These estuaries, with a total area of 1,312,000 acres,
may have their streamflow intercepted and diverted to cities and
irrigation enterprises along the entire coast if the Texas Basins project is completed. This grandiose scheme envisions a man-made
river paralleling the Gulf coast for 418 miles from the Sabine River
to the Lower Rio Grande valley.

The greatest of the Pacific coast estuaries is the Sacramento–San Joaquin estuary in California. This notable series of bays, tidal channels, and rivers extends many miles inland from the Golden Gate Bridge at San Francisco. Currents are present even in the "Delta," a reclaimed tidal marsh which receives the flow of the Sacramento and the San Joaquin rivers before they enter Suisun Bay, San Pablo Bay, and San Francisco Bay. The mixture of salt and fresh water extends for 50 miles from San Pablo Bay into the western Delta. Where the marshlands have not been claimed for agricultural use, they are managed as waterfowl habitat. In the network of tidal channels throughout these inland marshes, Chinook salmon, striped bass, American shad, white catfish, largemouth bass, and black crappie abound.

The tides and the waves are the sovereign forces of the seashore, shaping and reshaping the edge of the continent with their ceaseless motion. Year after year they nibble away cliffs and promontories, grind rocks into sands that form beaches and dunes, and carry the flotsam of the world to shore, where prowling beachcombers hunt for treasures. Beautiful shells, striking gemstones, grotesque pieces of driftwood and wondrous glass floats from far-off Japan are part of the water's bounty.

The tides, which are worldwide waves, and the near-shore waves are the greatest manifestations of the dynamics of the sea. The wind sets the waves in motion, and the gravitational forces of the sun and moon cause the tides. The centrifugal force of the earth's rotation, the configuration of the continental shelf, and the shape of the shore itself produce fascinating variations in the landscapes at the edge of the sea.

When the gravitational forces of the sun and moon are combined, the vertical rise and fall of the water is magnified. In the Bay of Fundy, tidal height is more than 40 feet. At Provincetown, Cape Cod, the tide rises approximately 9 feet. In Chesapeake Bay, the tides move as a series of slowly progressing waves up the length of the 150-mile estuary. Within this distance there may be two high-tide areas with a low-tide zone in between. The tidal range along the Gulf shore is small. There is usually one high and one low tide each day at Pensacola, Florida. Along the Pacific coast, tidal heights vary from a few feet to 14 feet at Balboa in the vicinity of the Panama Canal.

Most tides of the earth consist of two high tides and two low tides

Gay Point, Martha's Vineyard. The cliffs are of ancient clays with fossils of clamshells, crabs, whalebones, and the skeletons of mammals. Black sands contain magnetite beds but are not valuable as iron ore. The blue clay of Martha's Vineyard was formed when the ocean water was high, after the melting of the Ice Age glacier.

each day. The highest tides appear when the earth, sun, and moon are in a straight-line configuration. Then the gravitational forces are greatest, resulting in differences between high and low tides. The smallest differences occur when the earth, moon, and sun form a right angle and the two heavenly bodies tend to cancel each other's gravitational effects. Then the tides are neither very high nor very low. Tide tables should be consulted for local variations.

Collectors and observers of seashore life wait expectantly for extreme low tides, called *minus* tides, which expose great stretches of sandy bottom, rocky shores, and tide pools. Then animals and plants usually covered with water can be collected or studied. Clam diggers, fishermen, and picnickers commonly plan their trips to coincide with the low and minus tides. Tide tables, published by the United States Department of Commerce, Coast and Geodetic Survey, are available in most seashore localities.

The times at which high and low tides occur vary from day to day owing to the movement of the moon in its orbit while the earth is making its own daily revolution. The result is that high tides or low tides appear a little more than 12 hours apart. This progression of the tide times explains why low or high tides occur part of the time during the day and part of the time during the night.

Tides, which are the greatest of waves—lasting 12 hours and 25 minutes and spanning half the earth's circumference—have a dominant influence on creatures living in the intertidal zone. The habitats animals select in the zone are influenced by the relative time these creatures are exposed to air. Life forms on high rocks are covered with water only a few times during the year when the highest tides occur. Plants and animals near the farthest reach of the lowest tides are exposed to the air infrequently. Between these extremes are many kinds of exposure and habitats. These result in zonal distribution of living things along the seashore.

Tides also have their effects on salt marshes. When the tide runs high, you can boat up through the tidal creeks that meander through grasses, rushes, and sedges where birds, clams, and fish live largely unmolested by man or his works. When the tide goes down, you can float on water that carries nourishment to forms of life in the sea. The rhythmic behavior of the tide is the visible dynamic aspect of the marsh. It does not necessarily directly affect the herons, long-billed marsh wrens, ducks, crabs, and shrimp. But these animals do respond to the time of exposure to light, temperature, and depth of water. Thus the life of the marsh is tuned to the lunar rhythm and the ebb and flow of water from land and sea.

Waves are among the most dramatic features in the varied world of the seashore. If you watch waves on relatively calm days or in storms, in winter or in summer, you can see that their patterns are forever different. They vary in height, in their velocity, and in how they flow gently up long sandy beaches or attack headlands and rocky shores with appalling power.

I sometimes imagine that Proteus, who dwelt in the sea and pastured the sea calves of Neptune, still controls the waves and gives them their multitudinous forms. I remember how in mythological times he attempted to escape the use of his gift of prophecy by assuming in rapid succession the shapes of a lion, then a boar with lightning in his eyes, a stream, a flood, or a fire. Unless mortals could cling to him through all his changes, he would give no answer to their questions.

Now, when I look at the whimsical sea, it seems to be as changeable as were the many forms of Proteus. When the winds are calm and the waves are subdued, the fire of the sunset sparkles on their moving crests. In the darkness of night the billows thunder beyond the shore. Near at hand the grinding noises of rocks buffeted together by waves reminds one of growling lions. On gentle sandy shores, the waves advance and then retreat, forming broad floods or flowing back to the sea in braided streams.

When gales and hurricanes blow from the open ocean, the protean power of crashing waves is most apparent. The awesome weight of 20-foot breakers beats with a force of 6,000 pounds per square foot. Large waves move breakwaters and jetties weighing a million pounds. Still, the tiny limpets or the barnacles clustered on a stone are not dislodged. Their bodies are cone shaped or pointed like steep roofs and thus present small surfaces to the direct impact of the waves.

Waves create grotesque caverns and caves in vertical walls along the seashore. The Sea Lion Caves, south of Heceta Head on the Oregon coast, are one of the famous examples of this power of moving water. The spouting horn at Depot Bay also illustrates the work of the capricious sea. Here, the waves enter a tunnel with an opening in the ceiling. Enormous pressure forces the water upward in a spray that resembles the spouting of a gargantuan whale.

In the open sea, water merely revolves in a circle as a wave passes; it has no forward motion. You can observe this on a small pond by watching a floating leaf which rises and falls with each wave but makes no forward progress. The action of waves on the shore results from breakers formed by the motion of water on the upward-sloping ocean floor. The waves drag bottom, and the faster-moving crests topple landward with enormous force.

The mathematics of waves is well known. Breakers form and the wave breaks when its height exceeds one seventh of the wavelength, which is the distance between waves. Along the shore, the drag on the bottom slows down the front of the wave, and the water rolling up behind reaches a peak and curls over in a crest that rushes forward and falls in a cascade of foaming seawater.

The length of the wave determines how fast it will travel and the amount of time between successive crests. The amount of time between waves in this hemisphere varies from 5 to 10 seconds, depending on ocean storms and winds. South of the equator the interval between waves may reach 20 seconds. If the time between waves is 7 seconds, you can quickly calculate that more than 12,000 waves beat upon the shore in twenty-four hours. The water that piles up on the beach flows back in rip currents that stream seaward in broad sheets or in channels scoured in the beach sand. When we think of wave action in terms of thousands or millions of years, the number of stones and sand grains ground ever smaller and smaller with the shifting waves becomes incomprehensible. Because this carving process continues through time without end, nothing is permanent along the seashore. Great changes come when storm waves destroy beaches, fill harbors with sand, and project rocks weighing hundreds of pounds to the tops of lighthouses. The most deadly waves are the seismic sea waves (tsunamis), which travel at speeds of more than 400 miles per hour. Generated by earthquakes, volcanic eruptions, or enormous landslides beneath the sea, tsunamis can cause flooding of seashore towns, destruction of boats and piers, and human death.

When the ocean is in a gentle mood, one of the pleasures of repeated visits to the seashore is the opportunity to observe the changes, great and small, produced by wave action. Rill marks in the sand change as each wave recedes. Backwash patterns, where stones, driftwood, or even clam shells deflect the water, are never identical. Through the seasons, some beaches are built higher and wider, others are washed away, and some become rockstrewn where only sand formerly marked the shoreline. This ceaseless change, engineered by the tides and the waves, is part of the varied world of the seashore.

To be privileged to visit an unfamiliar part of the American seashore is a time to enjoy new vistas and learn the ways of different plants and animals. There is great biotic variety along the Atlantic, Gulf, and Pacific coasts. The visitor who travels from Louisiana to Labrador leaves a world of sand islands, marshes, and gleaming beaches and arrives in the realm of boreal woods that slope down from the Canadian provinces to sea level. In Maine, forests of spruces, white pines, hemlock, and balsam firs border rocky coasts once scraped clean of soil by the Ice Age glaciation. North of Maine, great colonies of seabirds nest in almost incomprehensible numbers on the precipitous cliffs of Nova Scotia and Newfoundland. Here there are no cabbage palms along the shore. Nor do any ghost crabs scuttle over white sands as they do on the beaches of Florida.

The visitor who journeys from southern California north to the shores of British Columbia also finds much variation in habitat along the way. In the south, wave action is powerful. Plants and animals of the seashore are impressively adapted to the breakers that explode against headlands and cliffs or flood into open bays and inlets. Along the northern coasts of Puget Sound and the quiet straits of British Columbia, giant forests descend to the water's edge and wave action is moderate. Here there is a limit to the number of zones of organisms that live in the narrow fringe between the tides and forests.

On Atlantic and Pacific shores the flora and fauna are distributed in three main groups, a northern, a central, and a southern group. As summarized by T. A. Stephenson and Anne Stephenson in *Life Between Tidemarks on Rocky Shores,* a few of the northern species found at Vancouver Island occur as far south as San Diego, California. The central group, with many species near Pacific Grove, overlaps the northern and southern groups. The southern group shows marked changes in the presence of species south of Point Conception and fades out in lower California as it approaches the tropical fauna.

Species on the Atlantic coast show a similar division in variety. The oyster, common from Cape Cod to Florida, also extends into the northern group. Most of the marine species in the northern group are best adapted to cold waters, but a few extend as far south as the Florida Keys. The northern region with its great abundance of fish attracts multitudes of seabirds, which nest on precipitous cliffs and rocky promontories.

The central group, centered about Cape Hatteras, includes species that are found in tropical waters. Many species live in the region from Cape Cod to Cape Canaveral. The populations of intertidal organisms seem to be adapted to intermediate temperatures. Those adapted to warmwater shores are characteristic of the coastal zone from Beaufort, North Carolina, to the Florida Keys.

The appeal of a specific coast depends on whether the visitor wants to see tumultuous waves, colorful landscapes, enchanting woods and seashore vegetation, or the diversity of biotic life. The distributions of flora and fauna are strongly influenced by temperature and by structural and physiological features which fit organisms for existence in different environments. In dividing the coastal zones into categories, scientists consider similarities in climates and types of biotic communities. Seasonal variations and circulation patterns of oceanic waters also are used to identify biogeographical regions.

The Boreal province with its cold waters includes the northwest coast of Alaska and has a fauna similar to that of Hudson Bay, southern Greenland, and Labrador. The Acadian subprovince extends southward to Cape Cod. The shoreland is rocky, glaciated, and subject to winter ice. The Maine coast, typical of this subprovince and with much allure for vacationers, is beautified by coastal woods, ragged peninsulas, inward-reaching arms of the sea, rocky offshore islands, and drowned river valleys. Although the coast is only 250 miles long as the gulls fly, Maine has over 3,000 miles of diversified shoreline. Large attached algal seaweeds, numerous mollusks, various crustacean orders, pelagic seabirds, and fish with bony skeletons and rayed fins live there. Many of these boreal species range far north into the Arctic.

The Virginian subprovince, extending from Cape Cod south to Cape Hatteras, is well known to multitudes of seacoast visitors. This stretch of land is classed by some scientists as a northward extension of the Carolinian province, which reaches south to Cape Canaveral and around the west coast of Florida to the Rio Grande in Texas. Some scientists call the northern coast of the Gulf of Mexico the Louisianian zone, since it is more tropical in nature than the coast of the South Atlantic states. In these zones, south of Cape Cod, sandy shores and extensive salt marshes predominate. Organic matter and microscopic green plants flushed from salt marshes and estuaries

Sitka spruce and Douglas fir trees come down to the edge of the sea in Oregon and Washington. Seedlings germinate on fallen tree trunks, which eventually decay and leave the new tree perched on enormous roots.

provide food for marine animals not found along the far northern coasts. The transition from the Acadian coast results in part from higher water temperatures, which inhibit cold-water species such as the large attached algae and the common barnacle. South of Delaware Bay, oysters become plentiful, as constant sand movement does not bury the rocky substrates where oysters attach themselves.

The southern tip of Florida is subtropical and supports a flora and fauna characteristic of the Caribbean. The low-lying shore is distinctly calcareous, with sands and coral reefs and a tropical vegetation. Here, mangrove trees with prop roots and tangled branches invade the sea. Innumerable gastropods provide seashells for the collector. On the coastal borders are hardwood trees, royal palms, ferns, orchids, and pinelands. Manatees, or sea cows, live in quiet waters, and black bears, cougars, otters, raccoons, foxes, and oppossums live in the hinterlands. Some of these, such as the raccoons, come down to the shore to forage on clams, turtle eggs, and fishes washed up by the waves. Some 300 species of birds constitute one of the most unusual bird faunas found anywhere along the American seashore.

The Pacific coast, in contrast with the Atlantic coast, has a high proportion of species that range far north and south and overlap the territories of species common to northern, central, and southern groups. The rocky shoreland from the Arctic to southern California is notable for its extensive algal communities, including the kelp beds that provide favorable habitats for sea otters and their principal food, the abalones and sea urchins. The seaside of both Oregon and California is generally mountainous, with rocky coasts, sand dunes, relatively small estuaries, and a general absence of large salt marshes and swamps. Along this coast, the visitor may see a menagerie of animals that varies from chitons, mollusks, and rockfish to spouting whales beyond the surf line.

Attentive observation of the biological forms that abound at the seashore can eventually lead to identification and enjoyment of hundreds of plants and animals. Even the amateur with little background in biology can learn to recognize trees, shrubs, and grasses adapted for life along the seashore. And the barnacles, crabs, jellyfish, clams, marine worms, sea anemones, starfish, beach fleas, gulls, ducks, herons, cormorants, sea lions, turtles, and inshore fish will be seen wherever they find favorable living conditions. They are a part of the varied world of the seashore.

2

Animals of the Seashore

MY FIRST CONTACT WITH AN ANIMAL of the sea came when I was a boy living on the prairie of eastern Nebraska. The animal was a queen conch shell that served as a doorstop in my grandfather's living room. It seemed to me a strange thing, unlike any of the animals I knew on our midwestern farm. It was a fascinating object to look at, handle, and hold to my ear to hear the "sound of the sea." At times it did reflect sounds that resembled the roar of the breakers I came to know so well in later years.

No one told me that this conch shell once was the home of a huge snail. I did not know then that it was one of the species of that large group, or phylum, of animals termed Mollusca that inhabit the oceans, brackish and fresh water, and the land about us. I still have the shell, which now is more than a hundred years old.

Not until years later, when I rode the Flagler Railroad from Miami to Key West, Florida, did I see their discarded shells piled up by the thousands after the edible bodies had been extracted for the market. I knew from my zoological studies in college that the animals in this group had soft unsegmented bodies, a rasping tongue like those of the snails and slugs that feed on vegetables in our gardens, and a muscular foot for attachment to and movement along the surfaces where the animals live. I also knew that the vividly colored protective shells were secreted by a bi-lobed flap called the mantle.

In my earlier simplified study of zoology in high school, I was presented with a malodorous preserved specimen of a clam worm, one of the genus *Nereis,* for dissection. My teacher, who knew only a little zoology, told me it was a segmented animal, one with many divisions of the body. To me it was a thousand-legged worm. Why he bought preserved worms for the class when he could have easily collected earthworms from the garden or leaches from the creek to illustrate the phylum annelida, I do not know. They would have been within our experience as young students.

Our teacher also failed to tell us that the segmented worms are a group of animals somewhere midway between the protozoans, those single-celled animals so numerous in the sea, and sea stars, sea urchins, sand dollars, and other highly developed animals without backbones. He missed a great opportunity to instruct us about the phylogeny of creatures in the sea and their evolution through the eons during which the ocean was the source of life for much of the whole animal kingdom.

The variety of seashells along the shore is almost unlimited. Collectors sometimes acquire hundreds of species after years of searching.

Through the years I learned more about clam worms. They make excellent fish bait, they stay alive on the hook for a long time, since they are so well adjusted to salt water, and they can bite fiercely with horny teeth embedded in the pharynx. I learned this early while using a tire iron to chip them out of their hiding places among the barnacles on Pacific shores. (Some fishermen have an easier way to collect them without destroying square yards of barnacles. They pour laundry bleach into a bucket of seawater and splash it on the barnacle beds. It brings the worms out in a hurry but does not destroy barnacles, limpets, and mussels. The practice, however, is illegal in some areas.)

Recently, I sat by a tide pool on the California shore near La Jolla where little caves and crevices adjoined the rock-strewn beach. At ebb tide the waves hardly disturbed the surface of the little pool among the reefs along the rocky shore. The sea stars were there. So were the chitons, clinging to the rocks among the algal jungles. The sea anemones were clustered in the mid-tide zone with bright colored tentacles spread, waiting to engulf unwary crabs that might come within their grasp. For the thousandth time I became aware of the magnificent array of life along the seashore. The thought occurred to me that much of the animal kingdom may be seen if one visits enough places along the seashore; we have only to avail ourselves of the opportunities that await us.

Let us look briefly at the animal kingdom on display along the seashore. The smallest animals are the protozoa, one-celled organisms or even some without cells. Protozoans are by far the most numerous creatures in the sea. They vary greatly, and because of this, scientists classify them into different groups under the kingdom Protista. Some 30,000 species have been recognized, but their individual numbers are almost beyond human comprehension. The single cells of most species, except large specimens such as paramecia, which some of us have studied in the zoology laboratory, are visible only under a microscope.

At times of superabundance they insult the nose, become luminescent on the moist sands of the beach, or change the color of the sea when the "red tide" blooms with Dinoflagellata. Dinoflagellates, which have two whiplike organs that enable them to spin and swim, directly concern humans since some of them produce a powerful toxin that accumulates in filter-feeding shellfish, such as mussels or clams. This substance is violently toxic to humans who eat the contaminated organisms.

Some of the protozoa synthesize their own food, as do other organisms of the plant kingdom, and are claimed as plants by the botanists. Other species eat organisms already present in seawater and are claimed as animals by the zoologists. But regardless of their food habits, these minute creatures are abundant everywhere in the sea and appear in many forms. Some with cilia (hairlike structures) grow on seaweeds or on shells. The importance of these multitudinous organisms is that they are the beginning of the food chain. Protozoa and very minute plants are eaten by very small animals. These, in turn, are eaten by larger organisms, including starfish, birds, whales, and man. If there were no protozoa, much of the life in the sea could not exist.

The sponges, phylum Porifera, are more advanced animals than the protozoa since they consist of many cells. These cells are assembled into bodies with shapes that vary from vaselike forms or colored

Almost every visitor to eastern shores finds the egg strings of the whelk cast upon the sand. However, they seldom see the big snail itself.

crusts on rocks to intricately branched and variously colored animals. Some sponges even resemble underwater plants.

The body of the sponge has many pores and chambers through which water containing food material is circulated by the beating of microscopic hairs. Living cells overlie the framework of the sponge, which consists of great numbers of microscopic glassy or limy spicules and organic fibers. These are the parts that remain in the common sponge we use in our bath or for washing the car.

Some species of sponge bore into abalone or other bivalve shells. Others live beneath layers of algal colonies. Sponges, in turn, are favorite habitats for certain species of shrimps, crabs, and barnacles. The ecological relationships between sponges and their associated animals are not well understood. Likewise, all the biological aspects of the sponges themselves are not well understood. Although they constitute the simplest group of multicellular animals, the amateur who decides to study them should be advised that they are a vague group of creatures which can be investigated scientifically only with the aid of chemicals, microscopes, and detailed descriptions and keys found in the scientific literature.

The next step up the scale of complexity is represented by the animals in the phylum Coelenterata. There are more than 9,000 species of animals in this phylum. Happily, many are recognized even by casual observers. Among the different classes of these animals are jellyfish, sea anemones, corals, and hydroids. The latter appear as soft, bushy or plantlike colonies on rocks in the intertidal zone. If you have a low-power microscope, you can sometimes collect hydroids and see them as tiny jellyfish-like creatures that swim. Other reproductive forms of the hydroids remain attached to rocks but can be scraped off to be examined under the microscope.

In spite of the many varieties, coelenterates have certain characteristics in common. The jellyfish, in particular, have stinging capsules, or nematocysts, which are capable of injecting poison into fish and even man. They have radial structure consisting essentially of a hollow sac surrounded by tentacles. And zoologically speaking, they have the ability to reproduce by budding in one generation and by sexual means in the next. The wide variation in form is exemplified by the corals, which form stony structures, and the by-the-wind sailor (*Velella velella*), an animal consisting of a colony of hydrozoans with a float and a sail. They are sometimes blown ashore by the thousands. However, they are not members of the group to which the Portuguese man-of-war belongs.

I have never been a worm enthusiast. But I went afield on the mud flats near Coos Bay, Oregon, with a marine biologist who made worms his most furious pursuit. He delighted in digging them up in slimy bay bottoms, scraping them off wet rocks, or screening them out of tide pools. He insisted that some are beautiful. I had to agree, especially when I saw their plumes waving like flower heads in the clear seawater. The worms, of course, are a vital link in the food chain of the sea and the seashore, since they convert innumerable creatures into worm flesh and in turn are converted into fish and bird and other kinds of flesh.

Zoologically speaking, the worms of sea and land, phylum Annelida, represent a significant advance in structure over the animals in the previously mentioned phyla. They present innumerable variations in size, habitat, reproductive process, and structure. They have heads and tails and right and left sides. They range in length from less than an inch to more than 100 feet. Three groups of worms are widely represented along the seashores of the world: flatworms, ribbon worms, and segmented worms.

Flatworms are abundant among seaweeds, on rocks, and in mud or sand along the seashore. Many are almost microscopic in size and feed on diatoms or other small animals. Flatworms also prey upon small clams, oysters, and barnacles. Some, such as the flukes and tapeworms, are parasitic on animals. Brightly colored flatworms sometimes may be seen swimming with undulating motions in tide pools.

Ribbon worms possess a proboscis used in capturing prey. These vary from an inch or less to more than 90 feet in length. Some are so fragile they break apart when handled, but the pieces may develop into new individuals.

Segmented worms, of which the familiar earthworm and the leech are examples, are members of a vast clan that includes several thousand species. All have bodies divided into segments bearing bristles that enable them to creep or swim. Some burrow into mud or sand; still others, such as the trumpet worms, live in self-constructed tubes.

If you dig for clams in shallow bays and estuaries, you may turn up one kind or another of the peanut worms. When it withdraws the ring of tentacles around the mouth it does somewhat resemble a peanut seed. These useful animals do in shallow marine waters what earthworms do on land. They eat detritus and swallow mud or sand to obtain organic material for food. The white peanut worm (*Sipunculus nudus*), is almost worldwide in distribution and may reach a length of 10 inches. Its striking appearance results from its iridescent white body.

The echiuroid worms, some of which are sausage shaped, include only about sixty species. Of special interest is the fat innkeeper worm (*Urechis caupo*), which occurs along the Pacific shore. Naturalists are fascinated by it because it lives in a U-shaped burrow and collects food by spinning a slime net which it periodically swallows. It is also of interest because it is associated with various commensals, including a fish, a scale worm, and a crab. Its life history will be described in more detail in chapter 5.

As we move up the scale of complexity in the animal kingdom we come to the Arthropoda, creatures with jointed legs. There are more than a million species of Arthropoda, of which three fourths are insects. Insects avoid the open sea with one exception, an ocean-going water strider that lives among the waves many miles from land. The other arthropods of the sea are a varied lot, including copepods, shrimps, water fleas, barnacles, crabs, and sea spiders. The sea spiders are strange creatures that appear to consist of nothing more than a half-inch or less of jointed legs clinging to eelgrass or to sedentary sea animals. Equally strange are the large horseshoe crabs, which are not true crabs at all, and lobsters.

Crustaceans are a large and diversified group of arthropods that have external skeletons, jointed legs that occur in pairs, and, in many species, bodies divided into head, thorax, and abdomen. In order to grow, they must periodically molt or shed their hard skeletons and then grow new larger ones. Among the most numerous of the Crustacea are the copepods, some of which resemble miniature shrimps less than one quarter of an inch in length. Many of these tiny creatures have a single red eye which reveals their presence in tide pools. Closely related to the copepods are the brine shrimps, which inhabit ponds of extremely salty water.

The barnacles, so common on rocky shores and pilings, are likely to stretch our imaginations when we learn they are crustaceans. But their life histories unmask their relationship with the scavenging rock lice, the beach hoppers that are flattened like dog fleas, the decapod crustaceans with ten pairs of legs, the spiny lobster, and the true crabs: the barnacle begins life as a nauplius larva, as a true crustacean should. Then it changes into a cypris, or bivalved larva, and finally attaches itself by its head to a rock, a boat, or a whale, and grows into a barnacle.

The spiny-skinned animals of the phylum Echinodermata are the familiar ones seen by most visitors to the seashore. Sea stars, or starfish, attract both adults and children with their five equal arms, their rough calcareous skeletons, and their mobile tube feet. Brittle stars (Ophiuroidea) resemble sea stars with their slender writhing arms with tube feet, which are used for feeding instead of traveling. The sea urchins, so common among the giant kelp forests, also have a five-part radial symmetry with movable spines attached to an internal shell. Sand dollars seen on the beach usually are the dead "shells" of these animals, which in life are covered with a velvetlike coat of grayish green spines.

The animals of the phylum Mollusca are the delight of collectors who prize the shells of clams, chitons, and the thousands of kinds of seashells found along the shores of the world's oceans. These animals are characterized by shells in an almost infinite variety of forms and by soft unsegmented bodies. Some have a large muscular foot. Many mollusks secrete their shells by means of a mantle which is part of the body wall. Mollusks are a diverse group of animals with elaborate feeding mechanisms. The group is represented by snails, sea slugs, bivalves (such as clams and oysters), mussels, scallops, and the octopus.

The top of the animal kingdom is the subphylum Vertebrata, the animals with backbones. These are the familiar ones of land and sea: sharks, rays, bony fishes, birds, and mammals. The fish are the most numerous, with nearly 7,000 species. Among the common mammals which come near or on the shore are sea otters, Steller's sea lions, harbor seals, elephant seals, whales, porpoises, and man. Man has invaded the domain of the sea with his diving gear, drained the seaside marshes, polluted the bays and estuaries, and cluttered the shoreline with shacks, houses, industrial plants, and highways. He is the most highly developed animal, if you are willing to accept his own opinion.

Sandy shores and long level beaches offer a delightful variety of inanimate and living things to find in the intertidal zone. The sea is always there beyond the tidal area. Inland are dunes, marshes, forests, rocky cliffs, and other landforms, all of which are inhabited by animals you can see and enjoy if you learn to explore their habitats and observe their ways.

The immediately visible animals are the birds that occupy the beach when the tide is low. Almost universally present are the pedestrian gulls patrolling the sands, from the filigree of froth left by receding waves to the jumbled masses of logs, oil barrels, shredded lumber, and flotsam from high winter tides. The active shorebirds—sandpipers, plovers, sanderlings, and willets—frequent the edge of the foaming water, while an occasional crow from the nearby forest makes flights in search of carcasses of small sea animals or morsels of food dropped by picnickers and beachcombing humans. All these birds are searching for the creatures that live among the sands of the beach or those cast ashore from the depths beyond the lowest tides.

Common starfishes come in various sizes and colors. They are rough, flat, soft, knobby or spiny, small or large, depending on the species. If an arm is lost the starfish grows a new one. Sometimes the lost arm grows into a new animal too. There is no way to preserve the color and texture of a starfish. Photographs are preferable for a day's record of finding them at the seashore.

The next animals you probably see are the beach fleas hopping out of the swash of algae left by the waves at the high-tide mark. These, and their other amphipod relatives, have long legs and bodies flattened from side to side. The large ones can leap several feet. They are active scavengers but most do their open hunting at night when the diurnal birds no longer prowl the beach. Those that live between the tides burrow deep into the sand until the tide recedes.

At low tide you may see thousands of holes in the sand. These can be nothing more than openings through which air has bubbled to the surface after a retreating wave, but if your footsteps cause a squirt of water from the hole, it may be the home of a clam withdrawing its siphons to lower levels. On the Washington coast just north of the Columbia River, I have traced hundreds of razor clams by the half-inch dimples they leave in the moist sand. Sometimes the hole, or dimple, shows the hiding place of the clam worm, that strange blue and green creature with so many bristlelike feet, which fishermen use for bait.

On the eastern coast, from Assateague Island and Cape Hatteras south to the Gulf coast, the ghost crabs can give you a merry chase on the beach if you try to catch one. They are animals of both land and sea: digging their burrows in the sand of the upper beach but periodically wetting their gill systems in sea water to obtain oxygen necessary for breathing. Most are sand colored; some are almost pure white, especially on the gleaming sands of the Gulf of Mexico. If you sit still on dark windy days, or carry a flashlight at night, you may see their periscope eyes in the entrances of their burrows and later see them scuttle across the sand and among the clumps of beach grass in search of food. But try to catch a ghost crab. It's like the old shell game: now you see it and now you don't. They simply vanish. Hence the name.

At the water's edge, watch for mole crabs, those little rounded crustaceans that frantically dig into the sand between waves. They live at the edge of the surf where they collect food in feathery antennae that serve as nets. They burrow quickly as the wave recedes and then reappear when the water washes over them again. The Atlantic coast species is *Emerita talpoidea* and the Pacific coast species is *E. analoga*. These little crabs, about the size and shape of a robin's egg, are used as bait by fishermen.

Sandy beaches are the temporary dwellings of various animals that do not live there the year round. In May and June on Atlantic shores, the horseshoe crabs crawl out of the sea at high tide, scoop out nests, and deposit thousands of eggs. Sperm from the males floats into the nests on waves, which then cover them with sand. The horseshoe crabs are more closely related to spiders and other arachnids than they are to the true crabs.

Many of the animal visitors on sandy shores come from the water. The waves sometimes strand fish in tide pools or leave them dead to be searched out by raccoons, gulls, and other scavengers. Skates are frequent victims of the tides, as are countless crustaceans and mollusks. Jellyfish, sand dollars, sea stars, and even whales are occasionally washed ashore.

The mole crab lives along the Atlantic and Pacific shores. This interesting creature tumbles about in the shallow water when waves run up the shore and then instantly digs into the sand. It has breathing antennae and feeding antennae, which strain plankton from the water. It has several pairs of legs, some of which propel the animal; others clean sand out of its gills. The eyes are on jointed stalks, which can be projected when the crab is covered with sand.

The real inhabitants of that narrow sandy land between the tides, however, are the animals that have solved the problem of living in a ceaselessly changing world by burrowing and hiding from the crushing force of the waves. They are animals that move constantly since there is little opportunity for permanent attachment in the shifting sands.

Beach populations of animals do vary from the highest to the lowest zone of the intertidal region. On the dry sands above the high-tide mark, ants make their nests, tiger beetles roam in search of prey, and beach hoppers jump in thousands as you cross the windrows of stranded seaweeds and flotsam from the ocean.

As you approach the upper wave limits, squiggly tracks, burrow entrances, and dimples in the sand reveal the presence of burrowing worms and amphipods (relatives of the beach hoppers) that dig below the surface for organic food. Also, look for clams of various kinds in the zone between high and low tides. At the water's edge and in the shallow sea beyond you find the world of sand shrimps, crabs, sea cucumbers, ocean perch, sand dollars, plumed worms, and crustaceans. These animals are adapted to life in shallow water and are never exposed to the drying effect of the wind-drifted sand.

Rocky shores are homes for many animals. Different habitats occur in zones that succeed one another from the lowest tide level to the tops of seaside cliffs. Many animals are adapted for life only in these rocky areas. Some find their best living conditions in rocky tide pools. Other creatures live in the zone above the high-tide mark where only the spray from large waves moistens the rocks. Still higher, far above the reach of the waves, seabirds nest in zones, one kind of bird above another. This selectivity of zones avoids competition for space and allows each kind to lay its eggs and raise its young.

On rocks between the lowest and highest tides along Pacific shores, acorn barnacles (*Balanus glandula*) are common animals that every visitor sees. These barnacles attach themselves to rocks where they are infrequently exposed at low tide. When covered with water they extend buff-colored appendages from their shells to catch food particles brought in by the waves. On rocks frequently exposed at low tide, great masses of the edible mussel (*Mytilus edulis*) are firmly attached by strong threads to rocks. These are relatives of the many species of clams, scallops, and oysters that flourish in various environments along the seacoast.

Farther south along the Pacific coast, the California mussel (*Mytilus californianus*) and the purple shore crab (*Hemigrapsus nudus*) are common. Where barnacles are abundant on rocky bottoms in tide pools you can nearly always see slow-moving colorful sea stars. Most sea stars have five arms, including the common star (*Pisaster ochraceus*), the spiny red star (*Hippasterias spinosa*), and the leather star (*Dermasterias imbricata*). The spectacular sunflower star (*Pycnopodia helianthoides*), however, has twenty or more arms. I once snagged one while fishing from the jetty at Winchester Bay, Oregon. It was purple and had a diameter of almost two feet.

Sea urchins are regular inhabitants in the rocky tide zone. These colorful "balls of spines" range from purple to green or red, depending on the species. The giant red urchin (*Strongylocentrotus franciscanus*) forms colonies among the rocks along the low-tide zone. Most sea urchins, however, live in water beyond the low-tide zone. They are especially abundant among the giant kelp plants.

Nearly everyone who explores the surf along rocky shores at low tide will encounter the beautiful flowerlike sea anemones. Although many of these creatures resemble plants, they are animal relatives of the jellyfish and other coelenterates, including the feathery-looking hydroids and the corals, which produce stony structures in the reefs in tropical seas.

The different species of sea anemones display many colors when their tentacles are extended. Some are brown to greenish; others are red, pink, yellowish-brown, or green; some show mixtures of white and pink. The tentacles possess nematocysts which discharge microscopic threads bearing toxic substances. These toxins aid in capture of crustaceans and small fish.

Some of the anemones likely to be encountered along the Pacific coast are the pile sea anemone (*Metridium senile*) with a long stem and tentacles at the top. The giant green sea anemone (*Anthopleura xanthogrammica*) grows on rocks in the surf along the Washington coast. The aggregate sea anemone (*Anthopleura elegantissima*) is common along the Pacific coast, where it lives among rocks and sand. When the tide recedes and it is temporarily exposed to air, it contracts into a ball of mixed tentacles and sand grains. The beautiful green and red sea anemone (*Tealia crassicornis*), found along California shores, also lives in places where it is frequently covered by sand.

A host of other animals live in the zone near and above the high-tide mark on shoreline rocks. Rock crabs watch from stony crevices.

Periwinkles, limpets, and rock lice live in spaces encrusted with blue-green algae. Still higher are rocks encrusted with lichens, while far above the turbulent sea, gulls, murres, guillemots, and other sea-birds occupy the stony walls, especially during the breeding season.

The animals of rocky Atlantic shores live in zones similar to those on rocky Pacific shores. Most of the species, however, are different. But the kelp plants, bladder wrack, rockweeds, blue-green algae, and lichens are present in zones with their associated animals. The barnacles, crabs, beach fleas, snails, horse mussels, nereid worms, sea stars, hermit crabs, brittle stars, and sea cucumbers are there. And along with them are herring gulls, ravens, and crows foraging for exposed animals and scavenging for dead creatures washed ashore by the waves.

Animals with shells appear in astounding numbers along the shore. Oysters, mussels, and barnacles grow in communities of thousands or millions where they are adapted to the turbulence of the sea. Some groups, such as the mollusks, occur in multitudinous variety, as any seashell collector can testify. Over 60,000 kinds of snails, periwinkles, and conches are found on land or in the sea. Some 10,000 species of two-shelled clams, mussels, scallops, and cockles have been named. All these have near or distant relatives which have unusual structures and live strange lives.

The chitons, for example, are strictly marine animals with eight plates on their backs and a foot, a head, a pair of gills, and a radula with teeth that rasp like a file when the animal is eating. The bivalves, animals with two shells, such as the clams, have a large foot for digging, and most have siphons or tubes which draw in and expel water containing minute organisms on which they feed. The tusk shells, open at both ends, include only a few hundred species. Some are found in mud and sand near the shore, but most live in the abyssal depths of the ocean. The strangest mollusks of all are the cephalopods, or squids and octopuses, with eyes, highly developed brains and nervous systems, and tentacles with suckers. More will be said in Chapter 5 about the fascinating life and habits of the octopus.

The mollusks, with their soft bodies and richly diversified shells, exemplify the abundance of life on land and in the sea. The phylum Mollusca includes more than 50,000 species worldwide. Anyone who walks the beach can see them or their shells. Common ones are the mussels, clinging to rocks and pilings by threadlike filaments.

The cone-shaped limpets, also abundant on rocks, are interesting because they return to the same spot after wandering in search of algal food.

Less conspicuous, but well known by people who dig, are the clams that live in mud or sand and extend their siphons to the surface. Many of these provide gourmet food for beach walkers. Among the choice species are the Pismo clams of California beaches, the razor clams of the Oregon and Washington coasts, and the Atlantic bay clams, including the cockle, the soft-shell clam, the butter clam, and the little-neck clam. To locate most of these, look for the telltale hole or dimple in the sand or mud.

To capture a horse clam, which also is called the empire, horse-neck, blue neck, blue, or gaper clam, you may have to dig a hole 1 2

A bag full of razor clams, and the fun of digging in the sandy beach, is ample compensation for dirty arms and knees. Some of the clams here are extending their feet and would soon bury themselves if left alone.

to 18 inches deep. Even more effort is required to capture the fabulous goeduck (pronounced "gooey duck") of the Northwest bays. It reaches a weight of 20 pounds and has a siphon which can extend as much as 24 inches from the shell to the surface. The valves of the shell are too small to enclose the animal, and the siphon extends from the shell from 6 to 12 inches even when retracted.

The univalves or "one-shelled" mollusks are of great interest to the collector because of their variety of color, design, and shape. These animals carry their shells on their backs and crawl on a strong mantle called the foot. The univalves have a head with eyes and tentacles for feeling, smelling, and tasting. The many families of univalves include the abalone of the Pacific coast and the snaillike periwinkle, which was introduced from Europe to the Atlantic coast some hundred years ago.

One of the largest shells is that of the Florida horse conch, which reaches a length of two feet. Among the most interesting shells are the wentletrap or staircase shells, found in all oceans and commonly along the Gulf coast of Texas. The white shells with ascending whorls are adorned with ridges formed as the outer lip of the shell is thickened. The New England wentletrap (*Epitonium novangliae*) has a reticulated pattern of spiral threads crossed by transverse lines. It ranges from Massachusetts and Virginia south to Brazil and can be found in beach drift.

The serious collector, of course, must go armed with a knowledge of how and where to collect and how to identify shell specimens. Numerous field books are available for local areas. A useful guide for shell collectors in the New York City area is *Shells from Cape Cod to Cape May* by Morris K. Jacobson and William K. Emerson. *A Field Guide to Shells of the Atlantic and Gulf Coasts and the West Indies* by Percy A. Morris lists 1,035 shells and enables the collector to identify shells of eastern North America, while *A Field Guide to Shells of the Pacific Coast and Hawaii,* by the same author, describes the western species. *How to Know the American Marine Shells* by R. Tucker Abbott is a guide to shells of both the Atlantic and Pacific coasts. And *Shells and Shores of Texas* by Jean Andrews is a treasury of information on history, ecology, and identification of mollusks of the entire northern coastal area of the Gulf of Mexico.

Visit the tide pools, if you wish, to see miniature zoos of strange and beautiful seashore animals. Their underwater worlds are easily accessible. Here, the creatures of the sea range from tiny, almost invisible plankton—food for inhabitants of the pools—to mussels, barnacles, crabs, limpets, sea stars, colorful worms, fish, such as the brilliant orange garibaldi, sea anemones, turban shells, and slow-moving chitons in somber shadows and dark crevices. Under rocks are worms that hide among algae or build tubes in mud or sand. If you move a rock you may be thrilled to find a small octopus and realize that you have revealed another niche of seashore life. Replace the rock so the octopus, tiny shrimp, snails, sea spiders, and other creatures can resume their ways of life.

The attraction of rock pools derives from their uniqueness. No two are the same. Life in the pools near low tide is abundant since the animals there are not exposed to drying winds and high water temperatures when the tide is out. In the colorful pools are coralline algae, feathery hydroids, sea hares, shrimp, sea slugs, and various kinds of crabs. In tide pools higher on the shore the fauna may be limited to barnacles, periwinkles, and a few browsing limpets.

A single tide pool may contain a spectacular display of a single species of animal. One that I watched for several hours in a rocky depression near Portland, Maine, was literally swarming with hermit crabs. They were engaged in a ceaseless scramble for sea snail shells, which they could use for larger quarters as the crabs outgrew their former homes. Houses for rent were at a premium, and the crawling, feeling, and testing of shells by the crabs made the underwater scene a fantasy of action in miniature.

If one sits patiently without moving, a tide pool can exemplify in miniature the remorseless exchange of energy that continues perpetually in the immensity of the sea. In the tide pool, as in the sea, the animals persist by eating and by being eaten. The mussels strain minute organisms through their siphon tubes to secure energy for maintenance, growth, and reproduction. The starfish envelops the mussel and, with its arms, pulls the valves apart and digests the succulent meat of the bivalve. Crabs snatch unwary fish or scuttle about for clams and bottom-dwelling worms. Shrimp flutter among the fronds of sea algae, catching tiny crustaceans which, in turn, are eating microscopic larvae and minute plants brought in by the tide. Some of this feeding is invisible. But for the larger animals, especially the fish, it is readily apparent. All the activity in the tide pool is a part of the web of life along the seashore.

On Atlantic shores, the crumb-of-bread sponge, an attached

animal, sometimes covers the bottoms of tide pools with a mat of conelike structures colored green by the algae that live within its tissues. In contrast, several species of shrimps swim actively in the higher tide pools. Some are transparent; others are opaque. On the surface a tiny insect, the seashore springtail (*Anurida maritima*), moves about like a water strider and feeds on dead animals and organic debris in the intertidal zone.

Because they are so attractive and easily accessible, creatures that live in their beautiful underwater tide-pool worlds should be observed but left undisturbed. Leave the plants and animals for other people to enjoy. Photograph them for later memories but do not take them to your home to die.

Anyone interested in fishes can find an unlimited variety along the seashore and in the sea. Some ichthyologists estimate that there may be as many as 30,000 kinds of fishes. Along any coastal area several hundred species may be found.

The different orders of fishes include descendants of primitive types that became extinct about 300 million years ago. Two of these are found along the northern Pacific coast, the Pacific hagfish and the Pacific lamprey. These jawless fish feed on other fishes, attaching themselves by means of a suction cup and rasping with horny teeth until they draw blood. Young lampreys, after a period of metamorphosis, migrate to the sea, where they become adults. Adult lampreys migrate into rivers to spawn. You can see them at the hydroelectric dams on the Columbia River, where they climb the fish ladders and attach themselves to the cement with their suction disks.

A dozen or more species of sharks, skates, and rays live along our coasts. These fish have cartilaginous skeletons, gill slits, and scales that give their hides a sandpaper texture. The deadliness of some of the larger sharks fascinate many people. The small ones, such as the Pacific dogfish (*Squalus acanthias*), are more common in near-shore waters. I have caught dogfish from jetties on the Oregon coast while fishing for rock bass and other bottom fish. A four-foot dogfish can break a nice casting rod if the retrieve is not handled carefully. The local fishermen at Winchester Bay make their own rods from two-inch-diameter bamboo with casting guides wrapped on with plumber's tape. These twelve-foot "telephone poles" will throw a heavy lure a terrific distance into the bay. And they have the spine to derrick a large fish or small shark up over the jetty rocks.

Small sharks are caught by local fishermen in bays and estuaries along the Carolina and Georgia coasts. This bull shark was caught at Cape Romaine National Wildlife Refuge. A common inshore species along the Gulf of Mexico, the bull shark even penetrates the fresh water of large rivers.

Most sharks give birth to live young. The skates, rays, and rat-fishes, flattened bottom fish with mouths located ventrally and adapted for feeding in sand or mud, on the other hand, produce egg cases that occasionally wash up on the beach. These may range up to twelve inches in length and contain up to a dozen eggs, which hatch into fully developed fish. These unique egg cases, also called mermaid's purses, possess tendrils which curl in at each corner. If you open one of these shiny leathery capsules you may find live baby fish inside.

Stingrays also are occasionally washed to shore. I have found numbers of these dead fish in the sand on Cape Hatteras. The Atlantic stingray, or stingaree (*Dasyatis sabina*), occurs from Chesapeake Bay to northern Mexico. Stingrays are plentiful in bays with sandy or muddy bottoms. The serrated spines on their tails are covered with a toxin that can severely wound a bather who accidentally steps on one. The wound is likely to become infected.

Of even greater interest are the electric rays, which are capable of giving potent electric shocks of thirty-five or more volts. The lesser electric ray (*Narcine brasiliensis*) occurs in the surf zone from North Carolina to Brazil. The Pacific electric ray (*Torpedo californica*) can knock a man down if he steps on one. Their maximum length is three feet. In nature the electric organs are probably used to repel predators and to stun prey.

The bony fishes are not often seen by those who walk along the seashore, but they are there in multitudinous variety, as any confirmed fisherman can testify. The fascinating grunion, which come ashore to lay and fertilize their eggs during the night-time high tide, are, of course, known by many people. So are the smelt that run up the Northwest coastal rivers in numbers beyond comprehension. On the Sandy River near Portland, Oregon, I have seen half a bucketful of smelt come up with one dip of a hand-held net.

Tide pools house many small fish, some of which are brilliantly colored. The garibaldi, mentioned earlier with its orange body, is more conspicuous than any goldfish. In contrast, the little blennies, gobies, and opaleyes are self-effacing and not likely to be seen unless you observe carefully. Some fish, such as the tide-pool woolly sculpin found on California coasts, lie motionless or crawl slowly over the bottom. The ocellated klipfish hides in rock weeds while the blennies hide under rocks.

Some fish of the intertidal zone come from the sea only at high tide. Those which are taken by angling or dipping include left-eyed and right-eyed flounders, surfperches, various rockfish and sculpins, gravel divers, pipefishes, and seahorses. Many of these are also found in estuaries and tidal marshes.

Fish of offshore Atlantic waters, lagoons, estuaries, and those that migrate from the open sea to spawn in rivers are too numerous for discussion in this book. In the northern coastal areas cod, pollock, haddock, and flounder comprise some 60 percent of the angler's total catch. Much of the remainder includes halibut, hake, cusk, tautog, mackerel, striped bass, smelt, and bluefin tuna. Permanent residents of the estuarine zone are smelt, tomcod, winter flounder, and smooth flounder. Anadromous fish, which ascend to rivers from the sea to spawn, include the Atlantic salmon, alewife, and blueback herring. In the southern stretches of the Atlantic coast, more than 500 species occur. These include barracudas, sharks, dolphins, snappers, sailfish, blue marlin, pompano, snook, tilefish and tarpon. It is a fisherman's paradise.

On the Texas and Louisiana coast, more than 400 species of fish have been counted. The variety of marine and estuarine environments here is greater than on any comparable length of coastline in the United States. Marshes, bayous, highly saline lagoons between the islands and coastal margin, the semitropical to tropical waters, and the habitats provided by turtle-grass beds and mangrove forests affect the lives of many fishes and permit assemblages of fish in bewildering variety.

For the angler, the naturalist, or the beach explorer who is interested in this engrossing world of aquatic denizens, *Fishes of the Gulf of Mexico—Texas, Louisiana and Adjacent Waters* by H. Dickson Hoese and Richard H. Moore is an indispensable guide. This book includes field identification guides to approximately 500 species of fishes and more than 600 photographs and drawings, of which 330 are in spectacular color.

Among the fascinating fish of the Pacific coast are the migratory species that spend a large part of their lives in the sea and then travel far up the rivers to spawn in fresh water. These include the sockeye salmon, silver or coho salmon, Chinook or king salmon, pink salmon, and chum or dog salmon. These fish find their way back to the rivers of their birth and even to the same spot where they were spawned. It is believed that each river has its own characteristic odor which guides the fish to their home grounds. The young fish, or fry, spend a few weeks to over a year in their migration back to salt water. The mature salmon die after spawning.

Two species that are favorites of anglers, the steelhead trout and the coastal cutthroat trout, do not die after spawning. They return to the sea and their offspring do likewise after one to three years. The steelhead gives a mighty thrill to the fisherman who hooks one. A shout goes up, "One on!" All fishermen along the river frantically reel in their lines and lures to give the "crazy man" running room as he wildly dashes over rocks, stumbles through campfires, and dashes into pools to keep pace with the mighty fish torpedoing downriver.

I used to enjoy a more sedate fishing for coastal cutthroat trout in Dairy Creek, a few miles west of Portland, Oregon. Each year the Oregon Fish Commission would dump a truckload of hatchery trout in the creek before the opening day of the fishing season. While many fishermen reveled in the sport of catching these small trout, I fished only in shallow riffles where the water cascaded over the stony bottom. This is where the cutthroat trout lived. Almost invariably I caught fish of superb eating quality that were twice the size of hatchery trout.

The Pacific coast fisherman has an enviable choice of surf fishing and near-shore angling. From rocky outcrops and jetties, the catch can vary from lingcod that weigh up to 100 pounds to the attractive striped seaperch that gives birth to live young. Kelp greenling inhabit shallow waters along rocky shores in kelp beds and near breakwaters. Black rockfish and cabezon also live among rocks. The rockfishes and scorpion fishes include more species than any other marine or freshwater fish family found along the California coast. Most are highly desirable for eating, and rockfish are important components of the sport catch from party boats.

The flatfishes constitute a fascinating group whose members are distributed from southern California to the Bering Sea. These include the starry flounder, lemon sole, rock sole, C-O sole, Dover sole, and mottled sanddab. Juvenile flatfish swim upright, as do other fishes. However, as they mature they undergo a metamorphosis until both eyes are on one side of the head and the fish swims on its side. It inhabits the bottom, where it adapts to the color of the mud or sand and is almost impossible to see when it is motionless. There

The lingcod makes good eating. These bottom fish can be caught by inshore boat fishing or by casting from rock jetties.

are right-eyed and left-eyed flatfish, characteristics which aid in their identification. As almost everyone knows, fillet of sole or flounder makes superb eating. Catching them in shallow bays and estuaries is also fun. When hooked they give one the feel of a flapping kite on a long string in a high wind.

The seashore has many mammal visitors. Some of these come down from the land to the shore. The raccoon regularly searches the marshes and sand beaches for crabs, turtle eggs, mussels, and other edibles in its varied diet. The mink is a common predator in seacoast marshes, dining on muskrats, nesting birds, and fish. The marsh rabbit is a denizen of brackish marshes, and the nutria is widely distributed throughout the Gulf coast marshes. The nutria was introduced by man to the islands off the Georgia coast but has recently been eradicated from the Blackbeard Island National Wildlife Refuge.

White-tailed deer come down to the dunes from seaside forests on Atlantic shores, and black-tailed deer make excursions to Pacific shore dunes from Washington, Oregon, and California forests. Even a black bear occasionally leaves its tracks along the beach as it searches for fish and other edibles cast up from the sea.

The bobcat, distributed across Canada and the entire United States, is a common inhabitant of wet swampy woodlands and an explorer of the tidelands. Many lesser mammals inhabit the dunes and coastal forests. The eastern harvest mouse, the striped skunk, the spotted skunk, and the cottontail rabbit are often seen along the seashore. The marsh rice rat inhabits southern marshes and swampy areas even on coastal islands. And at night the big brown bat flies over marshes in search of flying insects.

Some of the mammals that come up from the sea are conspicuous because of their size and numbers. Among these are the harbor seals and fur seals. The highly gregarious California sea lions are spectacular animals that breed along the Pacific coast from Mexico to the California Channel Islands. These mammals are conspicuously sexually dimorphic. The males, or mature bulls, weigh as much as 600 pounds; the adult females only weigh 100 to 200 pounds or more. On land the sea lions are more or less indifferent to man as the bulls protect their harems of numerous females and the pups wrestle with one another or play by swimming and riding the waves.

Nonbreeding groups of sea lions occur regularly over the eastern Pacific, ranging as far north as British Columbia. They are adept at

The raccoon is one of the great scavengers of the seashore, eating crabs, turtle eggs, small sea creatures, and wild plums in season. Its curiosity and traveling ability keep it well fed.

climbing over slippery rocks and can gallop for short distances over sandy beaches. They are great sleepers when they lie on sandy or rocky shores. The alarm cry of a gull or the near approach of a boat, however, may cause a general stampede into the water. I witnessed this once when I took a boat trip around the Three Arch Rocks along the Oregon shore. When we approached the middle arch the entire aggregation of mammals dashed into the water, and suddenly the sea was filled with barking beasts that constantly emerged chest high near our boat. We were in no danger, but their disturbance at our intrusion gave us an eerie feeling.

Other large mammals abundant in our coastal waters are the Steller's sea lions. Much larger than the California sea lions, Steller's bulls may weigh up to a ton. They breed on rocky shores as far north as Alaska. Another large mammal is the northern elephant seal, which frequents the outer Channel Islands and is characterized by a large inflatable proboscis that makes a loud warning sound. Cetaceans are marine mammals that do not come ashore. The largest of these are the killer whales, which gather in groups, or pods. These animals, with the distinctive high dorsal fins and white markings, are ferocious killers of whales, porpoises, and seals. But they do not seem to molest men who invade their domain with scuba gear.

One other mammal which does not come to land is the sea otter, which inhabits the kelp beds along the Pacific shore. These shy animals can be watched from the shore at Pebble Beach near Monterey and at other places where sea urchins, abalones, crabs, and various invertebrates provide an abundant food supply. This 3-to 5-foot-long swimmer has the finest fur of almost any animal and is now rigorously protected. It once attracted trappers and explorers and nearly became extinct.

The greatest of the sea mammals that come near the shore, of course, are the great whales that migrate thousands of miles from arctic to tropical waters. The California gray whale which passes southward along the coast of Oregon in October and November attracts many visitors in boats, especially at Los Angeles and San Diego, and later at their breeding area in the Sea of Cortez. It is a great tragedy when one of these huge mammals comes to shore and dies, and, unfortunately, this is not uncommon. The gray pilot whale occasionally strands in large numbers on the Georgia coast. Other whales sometimes come to Florida shores, and great efforts to help them back to the sea are made by admiring humans.

The penguinlike murres congregate in great colonies during the nesting season. They fish in the open sea where cold currents provide multitudes of fish.

3

Birds
Along the Seashore

IF YOU WERE TO CLASSIFY ME according to my ecological interests I suspect you would call me an "edge man," since I have lived mostly in transition areas between major ecosystems. I was born and lived my early years where the deciduous woods merged with the prairie. Later, I lived at the edge of Lake Michigan and felt most at home where the forests and dunes came down to the water's edge. In the years that followed I spent much time along the coasts of the continent. In all these ecotones, or blending zones between two physical and biological realms, I have been enchanted by the diversity of plants and animals that venture into these half-worlds and use both land and water for their homes. The best-adapted creatures are the mobile ones, particularly the birds.

Birds of one kind or another are always present along the seashore. The outstanding ones are the shorebirds, especially during migration periods. When the vast hordes have flown northward to the Arctic, numerous individuals remain to breed and raise their young on sand dunes, in marshes, and around saltwater bays and impoundments. Thus, even in summer, you can see willets, killdeers, and oyster catchers. Some are present from midwinter until late fall. And the gulls, of one species or another, are there the year round.

The remarkable variety of birds derives from the many environments along the coastline. Each topographic feature has its own ecological character which provides food, shelter, nesting space, and winter haven for birds with specific living requirements. Sanderlings are water-edge birds whose flocks ebb and flow, like animated mats, as they hunt for sustenance in the sands. Gulls raise their young in raucous colonies on lonely islands off the North Atlantic and the Pacific coasts. On these islands, with their storm-scarred rocks, the birds are safe since few mammal predators exist. Even seabirds such as Cassin's auklet, the ashy petrel, and the gannets must come to land if they are to reproduce and perpetuate their kind. With them on the nesting grounds are the gulls, cormorants, and oyster catchers. Although many of these species live in proximity with one another, each has its own requirements and preferences, including food. Some dive for food while some are scavengers; some are nocturnal, and others are diurnal.

The ecological wonderland of birds along the coastline becomes apparent when one studies their diversity in a limited area, such as one of the barrier islands of the Middle Atlantic coast. The Chincoteague National Wildlife Refuge in Maryland and Virginia, for example, contains less than 10,000 acres, consisting of sandy beaches, low dunes, salt marshes, pine and oak forests, ponds, and potholes, yet more than 250 different species of birds have been observed here. The list ranges from loons, grebes, cormorants, herons, ducks, geese, plovers, gulls, and terns to morning doves, hawks, owls, woodpeckers, flycatchers, wrens, thrushes, warblers, and sparrows.

On the Pacific coast, seabirds are abundant because of the rich food supplies of fish which, in turn, subsist on incomprehensible numbers of plankton in the nutrient-rich upwelling waters of the ocean. Here murres, puffins, auklets, guillemots, and cormorants dive for food while the omnipresent gulls feed on fish that come to the surface. Food is only a minor problem for these birds, but nesting space is at a premium since the small islands and sea stacks offer limited area for breeding and nesting.

The different species manage by utilizing slightly different habitats. The gulls nest in colonies of many thousands on protected grassy slopes. The cormorants sit in rows on barren rocks and raise their young on precipitous seawalls. The murres pack together in white-shirt-fronted choirs that protect their eggs from predator gulls. The auklets and puffins dig tunnels in which they breed and raise their young. This accommodation between species is one of the finest examples of ecological adjustment in a fiercely competitive natural community.

For greatest pleasure, the visitor to the seashore should do more than merely compile bird lists. Bird behavior is much more interesting. When you observe birds that live in different habitats in relation to their food, shelter, breeding habits, and physical attributes, their adaptations to life in the natural world become more meaningful. Even watching the evidence of bird intelligence affords opportunity for speculation about learning processes and acquired behavior.

Consider one example that gulls offer to watchers on Atlantic and Pacific shores. The herring gulls, western gulls, and glaucous-winged gulls all practice the art of dropping clams on rocks to break their shells. But the seagulls also drop clams on sandy beaches where the shells do not break. Their thinking processes thus appear to be limited.

On the other hand, gulls can recognize friends much as a dog recognizes his master. At five o'clock each afternoon, one of my friends at Oceanside, Oregon, used to toss scraps of bread and meat into the air at the edge of the bluff that separated his house from the sea far below. Even though no gulls were visible at the time, they would appear in minutes to dine on his food-laden patio table. When I tried to lure the gulls in this fashion they never came. Nor did they come when my wife tossed food into the air at the regular hour for feeding.

Observations of the instinctive behavior of birds along the seashore can be an enriching experience. Cliff-nesting birds, for example, breed in shifts as the calendar advances. Some congregate in great rookeries for mutual protection, and these colonial birds tend to breed simultaneously. This shortens the time they are prey to marauding mammals and plundering gulls, ravens, and crows. Gulls adjust their feeding times to the tides. And auklets and petrels change partners on the nest and do their fishing at night when danger from enemies is minimal.

There is much diversity in feeding habits among ducks. Some are dabblers, such as the green-headed mallard that prefers the shallow water of bay edges and marshes. The canvasback dives for its food and has it purloined by the American widgeon, which also steals food from the bills of coots. The American eider is the champion diver, descending to sixty feet for the mollusks and crustaceans which are its principal food.

The probing birds—curlews, tattlers, knots, dowitchers, and plovers—specialize in their food-gathering activities. The sanderlings and some of the sandpipers frequent the tide line and pick up small organisms that drift in and out with the surf. These birds work in flocks attuned to the ebb and flow of the wave currents and thus utilize a portion of the beach not frequented by other species. The short-billed plovers do their food gathering among the flotsam and sea wrack where flies, sand hoppers, and other organisms are abundant.

On extensive mud flats, when the tide is out, you can see the semipalmated, piping, Wilson's and black-bellied plovers searching for shellfish, sea worms, insects, and small crustaceans. Then, in marked contrast, you may see the black skimmer flying over the bay, skimming the water for small fish and other surface-swimming creatures. If you are on the West Coast, look for the black turnstone flipping rocks, driftwood, and seaweed clumps to see what lives beneath. Out beyond the waves the pelicans will be diving like animated rockets for fish near the surface. With them will be the cormorants diving and swimming to great depths for their prey. Three species of cormorant occur along the shore, but the pelagic cormorant and Brandt's cormorant prefer the salt water farther out while the double-crested cormorant can be found both along the coast and inland on bodies of fresh water.

Ducks also specialize in their habitats. The oldsquaw is an ocean bird, while the harlequin duck seeks its food among rocks in turbulent waters. The surf scoters raft on the open ocean regardless of the weather and seek their food from the bottom in the breaking surf. In contrast, the Virginia rail and the sora prefer the marshlands and bogs near the sea. They are seldom observed since their narrow chickenlike bodies enable them to slink through tangled swamp and marsh grasses in silence. The only evidence of their presence may be the grunting sounds of the Virginia rail and the whinny of the sora.

The sea-cliff birds specialize in two ways. The glaucous gull, Brandt's cormorant, and the common murre nest on rocky ledges, cliff tops and offshore islands. The rhinoceros auklet digs burrows in grassy areas on headlands and inaccessible sea stacks. The tufted puffin, a few of which nest on the California coast, prefers vertical sea cliffs. The common characteristic of these colonial birds is that they feed at sea, some at convenient flying distances and others far from land. Only the ocean could provide sufficient food for these congregations of breeding birds, which sometimes number in the millions.

Each year countless birds make the incredible journey up and down our seacoasts. More than half of all North American shorebirds leave their winter homes and return to the Arctic tundra for their breeding seasons. They do this in response to internal stimuli and the lengthening days of spring. The seashore provides an open route of travel and a dependable source of food for these avian travelers.

The shorebirds are among the great wanderers of the bird world. The best time to see them is in spring when they are returning to the far north to breed or in winter when they are flying south to sojourn along the Gulf coast. There are no specific times for their arrival and departure. Unlike the swallows that arrive punctually each spring at San Juan Capistrano, the migrating times of shorebirds, ducks, geese, gulls, and the birds of seaside forests vary annually according to temperature changes, storms, and food supplies.

The shorebirds belong to the order Charadriiformes, which also includes the gulls, terns, and auks. Suborders include the jacanas, oyster catchers, plovers and turnstones, sandpipers, avocets, and phalaropes. Some are long-legged, such as the stilts, others are short and squatty, and the sandpipers are small. Many shorebirds are dull in hue, but the distinctive color patterns of the killdeer with its two dark neck bands, the spotted sandpiper with its brown breast spots, and the wandering tattler with its yellow legs and dark bill longer than its head make these birds easy to identify.

Other clues for identifying birds include voices, behavior, and geographic location. The killdeers, curlews, and godwits have loud distinctive calls. The yellowlegs whistle. Some of the sandpipers utter shrill piping calls. The black oyster catcher opens shellfish and searches for limpets on rocks. Phalaropes spin or swim in circles to stir up small invertebrates. Sanderlings feed in flocks, while some of the larger shorebirds are more or less solitary as they probe in the mud with their long bills.

Migrant birds along the seashore differ greatly in their travel patterns. Some sanderlings cover 22,000 miles in the round trip from their Arctic summer homes to their wintering grounds in Patagonia. Others stop at the Gulf shore. The golden plover breeds in the Arctic tundra, moves to the coast of Labrador in fall, and then flies 2,000 miles over water toward eastern Brazil before moving inland to the pampas of Argentina. In spring, the plover flies northward to the shores of Venezuela, across the Gulf of Mexico, and up the Mississippi River to Canada. In contrast, the bays and marshes from New Jersey to North Carolina are free of brant geese until the first hint of winter brings them overnight in thousands. The next day they are there, tipping up like puddle ducks, feeding on eelgrass and sea lettuce. Sometimes the snow geese come with them.

Block Island, off the Rhode Island coast, and Cape Cod are exceptional places for finding birds after cold fronts have moved through. Birds that fail to follow the coast along New England shores are frequently blown off course and come to rest in fantastic numbers on these outer lands. On windy days, one may see shearwaters, petrels, and even a phalarope. The gulls, of course, are there by the hundreds, apparently unhindered by the blustery winds. In autumn, gannets may be seen on the wing, diving into the sea. Surf scoters buoyantly ride the ocean swells, while long lines of common eiders course swiftly over the waves.

On the mainland, bird migration along the shore is apparent from October to late spring. The tree sparrows arrive in Massachusetts by early November, pause for a while in the lowlands, and then move to higher wooded areas. They return to Labrador in early spring. Among other autumn birds are the brown creepers, dark-eyed juncos, golden-crowned kinglets, blue grosbeaks, black-billed cuckoos, and myrtle warblers. By mid-March, robins, phoebes, black ducks, and Canada geese will have arrived at Cape Cod on their way north.

The graceful gulls and their relatives, the terns and kittiwakes, are masters of the wind and are great wanderers along all our coasts. Gulls are adapted to environments that range from Arctic seas to high mountains, but they are preeminently birds of the seashore. You can hear the haunting cries of the glaucous-winged gulls above the booming surf of the Pacific and see the herring gulls in the garbage dumps along the Atlantic coast. The laughing gulls poach goodies from around your camp table on Hatteras Island. And the black-legged kittiwake wanders down the coast from Labrador or Newfoundland to southern New Jersey, or from Alaska south to Baja California along the Pacific coast.

If you are observant when you take the ferry out of Anacortes, Washington, or cross the Strait of Juan de Fuca, you may see Bonaparte's gull migrating between its Alaskan breeding ground and its winter home in Mexico. The California gull winters on the Pacific coast, but do not look for it there in summer; it breeds on inland lakes in California, Utah, Wyoming, and Canada. Heerman's gull breeds on islands in the Gulf of California and along the coast of Mexico and winters from Oregon to Guatemala. It is the only North American gull that wanders northward after the nesting season.

There are many other gulls along our shores. Along Atlantic shores herring gulls are among the most widely distributed over the northern hemisphere; as many as 200,000 scavenge in the garbage dumps in the vicinity of New York City. The great black-backed gull, the largest of all gulls, breeds from Nova Scotia to Labrador, and also along the Maine and Massachusetts coasts, and winters as far south as Florida. The laughing gull may be seen all along the Atlantic coast and west to Texas.

On the Pacific coast, the glaucous gull breeds in the far north but wanders south to California. The ring-billed gull winters along the coast from British Columbia to Mexico as well as along the Gulf coast from Texas to Florida. The large western gull nests on the Pacific coast from Washington to Baja California. The medium-sized mew gull—also a frequent visitor to city garbage dumps—nests along the seacoasts and inland along rivers. Look, too, for the least tern, the smallest of the North American terns, which breeds along the California and the Gulf and Atlantic coasts.

Gulls are intelligent birds and are cooperative among their own kind. They are adaptable opportunists with gullets that will accommodate stolen eggs, fish, crabs, barnacles, garbage, and the meat from dead seals. They are web-footed for swimming, have sturdy legs for walking, and long wings adapted for soaring and sailing. Few other birds possess these three capabilities.

A visit to a seabird colony is an unforgettable experience. One of the most spectacular of these is the Three Arch Rocks colony opposite the village of Oceanside, Oregon. Here, three great rocks rise as islands from the sea to forbidding heights. Through time, the waves have carved huge tunnels through each of these sea stacks. Since these islands are a bird refuge, one can enter only by permission from refuge headquarters—even a boat trip is dangerous because of submerged rocks in the billowing sea. A climb up the rocks is also forbidden because of the precipitous slopes and the thousands of nesting birds.

The last time I took a boat trip around the islands we were afraid to enter the arches with the boat for fear the waves would dash us against the tunnel walls and because of falling rocks dislodged by the birds. Adding to the bedlam were sea lions, which left their resting places and came voicing their defiance on all sides of our boat. High on the rocks, you could see as many as a third of a million murres, standing side by side like penguins. These birds do not build nests, but rather lay their eggs on the same stone ledges year after year. With them are gulls, petrels, cormorants, kittiwakes, and tufted puffins. The gulls nest there and they scavenge the eggs and young of other birds when the opportunity arises. The tufted puffins avoid most of the pushing, shoving, and egg predation since they nest in soil tunnels excavated between the rocks. When the sea is turbulent and all the birds are present, the population on this seventeen acres of rocks approaches a half million.

Some of the great seabird colonies on the North Atlantic coast are still relatively undisturbed, not exploited now as they once were by early fishermen and explorers. The great auks which once nested in the vicinity of Cape Cod are now extinct. These fat, flightless penguinlike birds were unable to climb to the higher rocks so they nested near the tide mark—easy prey for boatmen who clubbed them to death and salted them down for food for sailors. The last great auks were killed in Iceland in 1844.

The rookeries that now remain on the islands of Maine and the cliffs of Newfoundland have different communities of birds. Gulls predominate on some. There are gannetries along the northeast coast at Cape St. Mary's, Funk island, and Bonaventure Island off the Gaspé Peninsula. Other species include kittiwakes, fulmars, black guillemots, and murres. American eiders along the coast of Maine nest in the spruce woods back from the shore, while Leach's petrels dig tunnels in the soft earth among the spruces.

Gull nesting colonies occur in many places in the United States. The California gulls leave the coast to nest by the thousands on Anaho Island in Pyramid Lake, Nevada. Gulls occasionally nest on sand dunes, but those along the Pacific coast prefer the safety of islands. In 1955, ring-billed gulls were estimated to have a density of 2,000 nests per acre on a 20-acre area on Little Galloo Island in eastern Lake Ontario. The glaucous-winged gull breeds in a large colony on Mandarte Island, British Columbia. The ecology of this community has been studied by Kees Vermeer.

On the breeding grounds the birds are organized in spite of the

Double-crested cormorants commonly nest in trees and on rocky cliffs. These young birds are still downy, with no sign of feathers. They remind one of black goslings, except for their throaty faces and hooked bills.

chaos of numbers. The courtship ritual has many facets, including neck stretching, pecking, bill rubbing, and the ceremonious collecting of weeds and grass. Fighting and defense of nesting sites is the order of the day. The chicks are guarded by their parents to prevent their being struck dead and eaten by gulls in adjoining nests. Animal behavioralist Niko Tinbergen observed that eggshells are removed from the nest soon after the chicks are hatched. The white inner surface of the shell, in contrast with the mottled exterior, is more visible to predators. If the young gulls survive for six weeks or until they can fly, they stand a chance of returning to the colony to mate in future years.

A roll call of the birds along our seashores includes a great many species. The seashore offers open sandy beaches, dunes, mudflats, sheer cliffs, salt marshes, mangrove tangles, and both deciduous and coniferous woodlands for birds of differing ecological adaptations.

The rocky shores and oceanic islands off the coast of Maine are visited at one time or another by species ranging from birds of the open ocean to the sweet-singing passerine birds of the mainland. Here it is possible to watch the gannets diving headlong into the sea in search of fish. In winter, the great cormorants appear along the coast. On a Penobscot Bay boat trip you can see herring, great black-backed, and Bonaparte's gulls along with common eiders, black guillemots, and Arctic terns. In summer, phalaropes rest on the bay waters while in the woods the songs of Swainson's thrushes, white-throated sparrows, and various warblers are heard. You may not see Leach's storm petrel because it comes from the sea to its nesting tunnel only at night. But its nocturnal calls can be heard and its nesting colonies located by the musky odor of the underground burrows.

In the New York City region the salt flats of Jamaica Bay Refuge are frequented by many shorebirds including the northern, red, and Wilson's phalarope, the American avocet, the dunlin, and occasionally the buff-breasted sandpiper. In summer, the roll call includes grebes, bitterns, ducks, and rails which breed in the marshes. Piping plovers and least terns nest on sandy areas and brant appear among the eelgrass areas on the bays during the fall and winter seasons. In stormy weather in the Montauk region you should look for dovekies, which winter as far south as New Jersey. Also look for razorbills, thick-billed murres, Cory's shearwaters, and flocks of scoters.

A great mixture of birds occurs on Cape Cod since many migrants come here in season. Easterly storms bring dovekies, Leach's storm petrels, black-legged kittiwakes, eiders, common goldeneyes and oldsquaws to the bay at Provincetown. The spring migration brings many warblers into the forests. In summer, whimbrels and jaegers appear along with other shorebirds. Common, roseate, and least terns and the piping plover nest among the beach grasses on sandy areas. Sometimes the red knot, the largest of the sandpipers, may be seen on its migration from Greenland to southern South America. It makes the round trip of 20,000 miles each year.

South of Cape Cod, the great barrier islands, the lagoons, and the salt marshes are frequented by a notable variety of birds. In Chesapeake Bay, the pied-billed grebe "helldives" for eelgrass and wild celery, and the common loon sits low in the water between periodic dives to unknown depths. Many woodland birds spend the winter here. Daily, the brown-headed nuthatch can be seen going down tree trunks, while on nearby waters ducks and geese abound. The canvasbacks are here, too, but not in such numbers as they were before the market hunters killed them by the thousands for the fancy hotel restaurants in Baltimore.

The Maryland - Virginia shore and the barrier islands of the Middle Atlantic coast are outstanding for their sheer concentrations of birds. Here, geese, ducks, brant, and many shorebirds are present in and around the sandy beaches, low dunes, extensive salt marshes, numerous ponds and potholes, and forests of pine and oak. At Chincoteague National Wildlife Refuge, since 1943, more than 250 species of birds have been observed, including herons, hawks, rails, plovers, sandpipers, gulls, terns, owls, woodpeckers, swallows, wrens, warblers, sparrows, juncos, and some rare ones, such as the greater shearwater, parasitic jaeger, and snowy owl.

Recently, in late May, I took the Chesapeake Bay Bridge and Tunnel on the way to Hatteras Island to observe birds and seashore life. The trip across the bridge resulted in an unexpected encounter. At the curio shop, where seashells were for sale, I asked permission to photograph a large hermit crab. While holding the shell in my left hand and the camera in my right hand the crab emerged and clamped its claws on my fingers. As the crab and its shell were too valuable to drop, I got the picture at the price of a bloody hand. It

Horseshoe crabs climb sandy beaches on Atlantic shores to deposit their eggs. The one illustrated here is only the horny shell cast off when the creature molted to grow a new shell. These are not true crabs but are more closely related to the arachnids. They have lived in the sea for hundreds of millions of years.

required the help of a salesgirl to detach the crab and return it to its cage. After this memorable moment I was content to watch the seaside sparrows exploring the rocks at the end of the bridge.

When I arrived on Hatteras Island the wintering snow geese were already gone. But gadwalls, blue-winged teal, shovelers, American coots, and snowy egrets foraged at the edges of the marshes. I watched several willets exploring the sea wrack along the surf. A mighty storm then came out of the Atlantic and stopped my ornithological adventures. The lightning, the cannonading thunder, and the drenching rain continued through a good part of the night, but my tent, pitched on a small knoll, stood firm and kept my sleeping bag dry. In the bright morning sunshine the laughing gulls came in dozens for scraps from my camp table.

My recent visits to the barrier islands, beaches, mud flats, salt marshes, estuaries, and sounds of South Carolina and Georgia brought back memories of former adventures with those most spectacular of marsh birds, the herons. The great blue heron is a splendid creature that fishes miles away from the heronry. Years ago, Herbert E. Schwan and I drew the boundary for the Signal Hill Natural Area in the Nebraska Sandhills around the only heronry in that vast prairie region. The herons had to fly a round trip of 15 to 20 miles to fish in the Dismal River.

During another spring and summer I spent many hours in a tree blind on an island among the nesting herons. Occupying some 300 nests were great blue herons, black-crowned night herons, double-crested cormorants, snowy egrets, and common crows. The filth and stench in this heronry were unbelievable. I wore a wide-brimmed hat when I walked beneath the trees, as the young birds were experts at squirting their excrement over the edges of their nests.

The great blue heron eggs hatched one at a time on different days, resulting in four or five different-sized babies in each nest. From the blind I could reach into a nest, capture all five young herons, and bring them to the ground to be photographed. After I had returned them to the nest and remained quietly in the tree blind, the adult herons would come to feed fish to their babies.

The handsome black-crowned herons with pearl-gray backs and white breasts sat on their nests appearing to have no necks at all until they were disturbed by a movement in my blind. Then the ruffled feathers on their necks would extend upward a foot or more. The cormorant young soon learned to climb about in the tree branches like parrots. When captured, they inflicted painful bites on my forearms and more than once drew blood. The crows, of course, lived there as egg thieves. When I camped on the island at night, the great blue herons talked to one another in guttural tones while the night herons added their quawks and croaks to the noisy confusion in the colony.

Long-legged birds are legion along the coasts of South Carolina and Georgia. Great blue herons and little blue herons are permanent residents, while common egrets are summer and winter visitors. The great cormorant is a permanent resident on the estuaries and sounds. The common loon, the red-necked grebe, and the rare whistling swan can be seen in winter. On the mud flats and beaches are plovers, willets, and various gulls.

Pond and dune edges are homes for fish crows, long-billed marsh wrens, red-winged blackbirds, boat-tailed grackles, and seaside sparrows. Clapper rails clack in the marshes and black skimmers cruise low over the water. Wild turkeys explore the forests, while overhead bald eagles, black vultures, and ospreys patrol the sky.

In southern Florida and along the Gulf coast it is a pleasure to see the brown pelicans. I doubt if they will ever be as numerous as the white pelicans which reside on Pyramid Lake near Reno, Nevada. There, at the mouth of the Truckee River, I have walked to the edge of what appeared to be white acres of pelicans. They can be seen when the cui-ui, those huge suckerfish which are found nowhere else in the world, are spawning.

The brown pelicans have now nearly disappeared from Louisiana and Texas because of the residual effects of the pesticide DDT, which breaks down into DDE and upsets the calcium metabolism of egg shells. Fortunately, the ban on DDT and other pesticides seems to be allowing a comeback of brown pelicans both along the Gulf coast and on the Channel Islands off southern California.

Who cannot watch these sleek dive bombers without pleasure as they drop with wings bent, neck stretched, and head pointed to spear the water and scoop up a fish in the expandable pouch? They are imposing birds even when they perch on pilings or waddle at the edge of the foaming tide. The slow unison of their wingbeats sets them apart from all other birds that fly above the wave crests of the sea.

The Gulf coast, from Florida to Texas, has a great array of habitats for shorebirds, marsh birds, and tree-nesting species that obtain their food from waters where fish and invertebrates abound. The Florida Keys, for example, have hammocks, mangrove swamps, and mud flats with immense areas of open ocean and bay salt water. The great sawgrass river of the Everglades, the big cypress swamp, and the south coast mangroves are homes for "great white herons," now designated as a white phase of the great blue heron. These areas also have reddish egrets, roseate spoonbills, and remnant colonies of wood storks. Ospreys and bald eagles are here, but the Everglade kite may not survive much longer owing to timber cutting and the draining of its habitat.

The Louisiana shoreline—6,952 miles in length if one counts all the indentations formed by river mouths, bays, and sounds—is an unexcelled habitat for gulls, terns, shorebirds, ducks, and geese. The brown pelican, once a permanent resident, has now been transplanted from Florida. The great blue heron, the Louisiana heron with dark neck and white belly, the snowy egret, and the white-faced ibis are common in the salt marshes.

Shorebird abundance in Louisiana varies with the seasons. The piping plover appears on migration in late summer and again in early spring. The semipalmated plover frequents the sandy beaches and mud flats somewhat later in autumn and in late spring. The long-billed curlew is numerous in fall and spring on sandy beaches. But I best remember its thrilling call, *cur-lee, cur-lee,* at its breeding grounds on the Great Plains where I did range research for many years. Even the willets along the shore with their startling loud cries remind me of the curlews on the summer prairies.

Gulls and terns are a numerous clan along the Gulf coast. Only the laughing gull breeds here, but migratory gulls commonly seen include the herring gull, ring-billed gull, and Bonaparte's gull. The terns are populous along the coast, and seven species breed within Louisiana. Forster's tern, the common tern, the roseate tern, and the least tern are familiar birds. The sooty tern, which breeds only in the Dry Tortugas, ranges widely over the ocean but is blown inland by tropical storms and hurricanes.

The offshore islands of the Gulf coast, with their sandy beaches, grassy flatlands, and tropical shrub lands, are attractive to many birds; nearly 300 species have been recorded at Padre Island off the coast of Texas, including the usual knots, sanderlings, avocets, and willets, as well as great numbers of transients during the migratory seasons. It is a mistake, however, to expect to find multitudes of birds near the shore as they make the northward flight over the Gulf of Mexico. Even the sparrows, hummingbirds, swallows, wrens, and warblers, which have migrated at night, continue inland, sometimes for many miles, where they may find insect and other food to replenish their depleted energy.

There is one bird, the seaside sparrow, that truly deserves its name. This elusive creature ranges from New England to Texas but restricts its territory to salt marshes and the wind-blown grasses near the surf. It has somber gray and dark olive streaks with a yellow spot between its bill and its eye. Its massive bill is adapted for

"Great white herons," great egrets, and other long-legged waders are common in Florida and other southern marshes. The great white heron is now regarded as a white form of the great blue heron. It has greenish-yellow legs. The great egret has black legs.

eating the snails, crustaceans, and small marine animals that it picks from debris washed up by the waves. The seaside sparrows prefer life near the ground. They scuttle across the strand or vanish among the grass clumps like ghost crabs.

The birds of the Pacific coast provide even a beginning or amateur bird watcher a pleasing diversity at any season of the year. The numerous environments vary from the warm coast of Baja California to the stormy winter habitats of British Columbia. In this long stretch of seacoast are sandy shores, mud flats, rocky cliffs, estuaries, sheltered bays, and offshore islands and sea stacks. The Pacific shore also is a migration route where avian visitors come from the far north and where winter visitors pause for rest and feeding in preparation for breeding inland or for the return journey to Arctic summer lands.

The coastal birds represent many families, but the ubiquitous gulls are the symbols of the seashore. Their cries can be heard above the thunder of the waves. No one can forget them once they have been heard on wild beaches or above the stormy sea. Thirteen species are known in the province of British Columbia. Around San Francisco, eight species are regularly seen, but only the big western gull, with black wings and back and white body and head, lives in the Bay area the year round. It nests at Point Reyes and on the Farallone Islands, where it commutes 25 miles daily to the city dumps and beaches for garbage, inshore fish, and clams along the beach. The Oregon glaucous-winged gulls do even better. They fly 75 miles or more from the bays and beaches near Tillamook and Astoria to the city dumps in Portland. Some of them cruise up and down the Columbia River and others fly directly over the Coast Range in western Oregon.

Many of the gulls are difficult to identify because of their plumage changes. Some of these take three years before the adult and sexual maturity stages are reached. The glaucous-winged, western, herring, and Thayer's gulls are large birds and are common winter residents. The smaller ring-billed gull is easily distinguished by the complete black ring around its yellow bill. The mew gull is the smallest of the gull family, measuring only 17 inches in length.

Bonaparte's gull and Franklin's gull are regular migrants. The Franklin's gull differs from the Bonaparte's gull because of its darker mantle and less white on the wings. Both have black heads in adult summer plumage. Heermann's gull with its white head, red bill, and black tail tipped with white is easily recognized. It specializes in robbing brown pelicans of their catch.

Among the birds that associate with gulls are the parasitic, pomarine, and long-tailed jaegers. These birds have projecting central tail feathers, white on the primary feathers, and hawklike flight. They are adept at stealing food from gulls while both birds are on the wing. If you are lucky you may sea a skua, a large, stocky, dark gull-like bird of the open ocean that occasionally visits the Pacific coast from its breeding grounds in Antarctica. It is an accomplished scavenger, preys on young gulls, and forces mature gulls to drop or even disgorge their food. Robert Cusman Murphy, an ornithologist of the American Museum of Natural History, states that it will viciously attack dogs or men when they approach its nest.

There are enough seabirds along the Pacific shore to satisfy even the most avid beach stroller. The gulls, cormorants, murres, puffins, ducks, and sandpipers come and go with the winds and the seasons. Sometimes they are here, sometimes there, but they are always pre-

Sanderlings feeding at the edge of the waves in Monterey Bay, California.

sent somewhere along the coast. The greatest pleasure comes when you can observe some facet of their lives in addition to just giving them a name.

The dunlins, for example, are common wintering birds that fly in groups of hundreds, somewhat after the fashion of blackbirds. Dunlin flocks sometimes contain astounding numbers. In winter, when thousands of these birds assemble on tidal flats and shore margins, they resemble moving carpets.

Many of the offshore rocks and islands are off limits to human visitors simply to protect the huge colonies of nesting birds. But someday, if you are climbing a steep rocky cliff along the shore, you may hear a thin piping whistle above the roar of the sea. If you climb slowly, you may discover the whistler and it will probably be a pigeon guillemot. It may be standing by its dark spotted eggs in a rock crevice or on the front porch of its nest, which is behind it in an earthen hole. It may even continue its whistling as you stare at its body, which is black with a white patch on each wing, and its bright red feet and legs.

You will record another memory of the seashore if you are watching with binoculars or if you are in a boat trolling for fish beyond the breakers. In late summer the shearwaters will be migrating or summering far from their breeding grounds in New Zealand and Australia. They may be swimming in single file, like a long undulating chain of beads, formed by hundreds of birds head to tail. If they are on the wing they will seem to "shear" the water as they fly in perfect formation above the azure sea. They will be a lovely addition to your memory of the marine picture.

4

Botany for Seashore Wanderers

I RECENTLY EXPLORED SOME OF THE DUNE COUNTRY along the North Carolina shore where the strand consists of almost pure sand and the wind drifts salt spray landward from the ocean. Heat and light reflected from the straw-colored substratum. The habitat was utterly devoid of vegetation.

This area, like other sandy beaches and dunes, was unpopulated with vegetation since plants have little chance of survival where the soil is being constantly shifted about by water and wind. When plant life does survive by the edge of the sea, it usually consists of annual species which grow in small accumulations of sand or soil and are adapted to a brief existence only in a salty environment. Such plants are seldom seen inland.

As a botanist, I found this habitat thought-provoking and speculated pleasantly on the influence of the sea on shorelines. I was reminded of contrasting shorelines along the rockbound edge of Maine, the saltwater marshes of Louisiana, and the plant-covered cliffs of Oregon, where luxuriant land vegetation comes down to the edge of the high-tide mark. But here on the North Carolina shore was the story of plants on sand flats above high tide and the evidence that grasses are important in dune formation. Actually, I had observed this story in many other places, such as the dune areas on Cape Cod, on Cape Hatteras, on Florida coastal islands, on California's sand strands, and on Oregon's dune mountains north of Coos Bay.

Here, on North Carolina's sand flats just above high tide, I found a sea oats (*Uniola paniculata*) plant making a tiny mound of sand behind its little leaf cluster, its roots permeating the sand in a binding network. To the leeward of its clump, new leaves were pushing upward and expanding the size of the plant. By capturing the sand it was actually making a small dune. Other sea-grass plants nearby were doing the same, and a few had captured several square yards of sand. The large panicles of sea oats that had matured waved with a golden glow against the blue sky.

Sea oats is so important as a sand binder that it is now protected by law on southern coasts. Formerly, it was gathered and sold for flower decorations, and the harvest was so intense that stabilized

Beach grass, the great sand binder on foredunes, holds the sand with long scaly rhizomes and numerous roots. It was introduced from Europe.

dunes began to move again. The more northerly sand grass (*Ammophila*) on Virginia, Cape Cod, and Pacific shores is less pleasing ornamentally, but admirably useful in stabilizing drifting sands of dunes that would otherwise move and cover everything in their paths.

The dunes near the seashore are excellent places to see and study pioneer plants. On the North Carolina sand flats I found the sea elder (*Iva imbricata*), a low shrub two to three feet in height. Its leaves are fleshy, as are those of many desert plants that are adapted to conserve water in arid environments. This member of the composite family grows on sandy coasts from Virginia southward. Another little plant associated with the sea elder is the prostrate dune spurge (*Euphorbia polygonifolia*). This annual can be recognized by its branches, which radiate from a common root, and by its milky juice. The cocklebur (*Xanthium echinatum*), an inhabitant of sandy beaches from Maine to North Carolina, is also here. This plant's burrs are plumper and its prickles are much shorter than the diameter of its body.

On the landward side of the dunes and inland from the sea, I observed that plants became more numerous, especially where the sands were stabilized and long-lived species were allowed to maintain themselves over periods of years. These plants had no need for adaptation to salt water, since the rainfall keeps underground saline water pushed back toward the sea. The gradual change in vegetation one sees on sandy lands as one walks away from the sea, and the succession of plant communities from nearly bare beaches to forests, is discussed in greater detail in Chapter 6.

The seashore botanist, if he or she lives in one place for a number of years, may learn to think of favorite plants in terms of the calendar, as many of us do in forested country inland or on the prairies of the Midwest. Seaside strawberries bloom in spring, the grasses and spurges bloom in summer, and the goldenrods bloom in autumn. But the vernal, or spring, aspect is not as conspicuous as it is on pastures and prairies, where the whole landscape bursts forth with variegated carpets of colorful flowers. In summer, the showy herbs are overtopped by prairie grasses, or the hepaticas and bloodroots in the forests are obscured by tall herbs and shaded by leafy tree canopies. In autumn, in woods or prairies, the yellows, whites, and purples of tall asters, the silver of beardgrass, and the orange of butterfly weeds make the landscape riotous with color. Not so on the seashore. On cliffs and high benchlands above the sea, the profusion of plants and their variegated patterns are there. But near the beach, the standing cover of grasses, forbs, and shrubs is usually distinctive only in conformity with local topography, soil characteristics, wind direction, and climatic factors.

In spite of the bareness of the beach we should remember that there are plants in the sea, but they are different in many ways from the land plants we know. The smallest of these sea plants—the diatoms—are more numerous than any other plants in the world. Diatoms are single-celled microscopic plants, with walls of a pectinlike substance that are covered with a layer of hydrated silica. These organisms, which make up much of the phytoplankton of the sea, are photosynthetic producing energy from sunlight. Along with the single-celled dinoflagellates, motile microscopic plants, they are the first step in the food web of the sea.

Diatoms have been called the "grasses of the sea" because they are grazed by a multitude of animals. These minute algae are eaten by small crustaceans, such as copepods, and are filtered from seawater by worms and a host of other small animals. These, in turn, are eaten by small fish; larger fish feed upon the small fish; these are used for food by yet larger fish, octopuses, eels, seals, and birds, and finally some are used for food by man.

Much of the plant life of the seashore, at the water's edge and in the sea, consists of algae, which are primitive types of plants. Botanists group algae by color: blue, green, red, brown, and blue-green. The green kinds grow closest to the shore. A common algae is sea lettuce, which you can see in tide pools and in water-filled channels between rocks when the tide goes out. On rocks above the splash zone of the waves, algae form mats which are grazed by snails, limpets, and other crawling creatures. In contrast, the giant kelp form beds a few hundred yards offshore. Some kelp plants exceed 100 feet in length and make veritable sea forests. The flat furrowed leaves of these giant algae are held aloft by small bulbs, or floats, filled with gas. The tubular stems are attached to rocks on the sea bottom by holdfasts, structures that resemble roots but are not roots. The kelps, or brown seaweeds, are frequently washed ashore during stormy seasons and make interesting finds for the wandering beachcomber.

There are hundreds of species of algae along our coasts. Some are shortlived and thus not easily noticed. Many have alternate stages of reproduction, which makes them look like different plants during their life cycles. Such algae are difficult to identify if you are not a professional botanist.

The variety among algae is endless. Some are free-floating. Many of the blue-green species live in colonies, forming scummy growths on wet soil, rocks, marshy bottoms, and in pools. Some of the green algae produce fuzzy, stringy growths, while others appear as thin lettucelike sheets (*Ulva* and *Monostroma*) or grow in the form of hollow tubes (*Enteromorpha*) that resemble intestines.

The brown algae predominate in the intertidal zone and in the ocean beyond. Their golden brown or olive coloration is produced by the pigment fucoxanthin in addition to the chlorophyll found in all living plants. The kelps belong in this group. But the variation in this phylum of algae includes plants that form black crusts on rocks,

Various species of *Fucus* and other rock-dwelling algae in the intertidal zone are characterized by dichotomous branching. Most are cartilaginous, flattened, and some show conspicuous midribs and swollen receptacles. They are slick and one can easily fall when walking on them.

grow in feathery plumes, exist as hollow sacs, or develop into brain-like masses.

In addition to chlorophyll the red algae possess a pigment, phycoerythrin, which absorbs the blue portion of the spectrum most intensely. Since the short wavelength of this light penetrates water deeper than light of longer wavelengths, the red algae can thrive in subtidal depths. Like algae of other groups, the red algae exhibit many colors and forms of growth. These vary from simple filaments to sheets and branched structures to corallike formations.

Along New England shores, Irish sea moss (*Chondrus crispus*) is one of the well-known red algae. It can be boiled and made into a pudding with milk, fruit, and vanilla flavoring. In former times, Irish sea moss was used to treat urinary infections and diarrhea, but it has many other uses because of its carrageenan content. Carrageenan adds body to ice cream and is used in beer, shaving cream, canned meats, water paints, ink, and laxatives. Boatmen can collect hundreds of pounds of this algae in a few hours by raking the clusters from the ocean bottom or picking them by hand from rocks. Irish sea moss sells for three to three and a half cents per pound.

One of the algae found most commonly along the Atlantic coast is rockweed (*Fucus vesiculosis*). This plant has a leathery mucilaginous structure and is very slick. It clings to rocks and pilings by means of holdfasts and is easily identified by its dichotomous, or forked, branching—flat fronds with pairs of air vessels on each side of the midrib. One should exercise care when walking on rocks covered by rockweeds. Beachwalkers may find rockweed from all depths that has been torn off during storms and washed ashore by waves.

The kelp *Laminaria agardhii* commonly grows on pilings. Others, such as winged kelp (*Alaria esculenta*) and sea colander (*Agarum cribosum*), are large, cold-water species. Sargassum (*Sargassum filipendula*) is characterized by bladders that grow on stemlike structures attached to the leaflike fronds. All these are members of the brown algae. Dulse (*Rhodymenia palmata*), as the species name indicates, is shaped like a large flat hand attached by a holdfast at the "wrist." It is deep red and commonly grows in the intertidal zone.

The giant kelp (*Macrocystis pyrifera*) that grows in massive beds off the California and Baja California coasts is the most impressive of all the algae. Specimens sometimes reach a length of 300 feet. These kelps produce large straplike fronds that sway like flat ribbons in the waves. The stems are attached by strong holdfasts at great depths in the ocean.

Scuba divers find a great variety of animals, including sea urchins and sea otters, in these underwater jungles. But there is danger here for human divers, especially if they become ensnared in nylon fishing line that has drifted among the fronds. My son, Donald, who has an international license to teach scuba diving, recommends that his students carry knives attached both to their ankles and arms so they can extricate themselves if they become entangled in a kelp forest.

The giant kelp is an important commercial resource. Large boats now harvest thousands of tons of kelp by means of mowing devices and conveyor belts. While it used to be harvested for their iodine and potash content, kelp is now used to produce alginic acid and its derivatives. Sodium alginate, for example, is used as a fabric stiffener. Alginates are also used in pharmaceutical products, dental impression materials, as stabilizers for ice cream, and as fillings and glazings for bakery products.

The kelp industry has become so important that plans for farming algae in the ocean are now being considered. The farms would be supplied with spores that are grown on plastic lines in coastal nurseries and then towed by divers to the sea farm and attached to the bottom. The giant kelp grows about 15 inches a day, and it is estimated that 4,400 square yards could yield 495 tons of harvested kelp per year. The crop could be used for food, fuel, and the production of methane gas.

One of the largest brown algae on the Pacific coast from California to Alaska is the bladder kelp (*Nereocystis luetkeana*). It is frequently cast up on shore by storm-driven waves. From a holdfast of rootlike structures, a long cylindrical stipe up to 75 feet long rises to a spherical float some 5 or 6 inches in diameter. Two groups of flat divided blades arise from this float. Each cluster consists of many blades, which may each be 10 to 15 feet long.

Another common species of brown algae is *Macrocystis integrifolia*. Often found on the beach, this species may be recognized by its numerous leaflike branches attached to long erect stipes that grow from a horizontal rhizome along the bottom of the sea. A float grows from the stipe where each blade is attached. The 8- to 10-inch-long blades have small teeth along the margins.

An alga of unique appearance is the sea palm, *Postelsia palmaeformis*. As its specific name indicates, it resembles a small palm tree, two feet or more in height, growing from rocks exposed to heavy surf. It seems miraculous that the hollow stipe can hold the many fronds upright among thundering waves. The plant remains upright and exposed to the air when the tide is out and is submerged when the tide is in. This species occurs from British Columbia to California.

There are so many other algae in the sea that a thorough knowledge of this bizarre group must be left to the specialist. As noted previously, many of the seaweeds have alternate stages of development and, since some of these are microscopic, they go unnoticed by the layman. With local handbooks, however, the observant person can identify at least some of the common species, especially the large and conspicuous ones.

Only a few marine plants produce seeds. Most are grasses or grasslike plants adapted to a salty environment and capable of reproducing themselves by rhizomes or by branching underwater stems that send up new leaves. The sea grasses are of extreme importance since they supply food and attachments for many sea animals.

These marine plants are grazed by birds, fish, and many invertebrates. Their decomposed remains add detritus and nutrients to the water. They stabilize bottom sediments, protect small animals from predatory fish, and preserve the lives of estuarine scallops and other animals that must find attachment during their larval stages. Not least of all they serve as nursery areas for fish, crabs, and shrimps and as homes for eels, flounders, sea slugs, turtles, innumerable diatoms, small algae, hydroid colonies, and other life forms that are indispensable links in the marine food chain.

Eelgrass (*Zostera marina*) is one of the predominant submerged sea grasses in temperate estuaries. Actually, eelgrass is not a grass. Instead, it belongs to the freshwater pondweed family. It and a few other flowering marine plants are called "sea grasses" because of their resemblance to true grasses that grow in swamps and on dry land. Most of these plants possess creeping rhizomes, which produce roots and erect shoots and leaves. Some reproduce sexually by male and female flowers. Pollen is transferred by water instead of wind or insects. The seeds tolerate and germinate in salt water.

Eelgrass grows in South Oyster Bay, Long Island, in the waters of Cape Cod, and elsewhere along the Atlantic coast at depths of 6 to 8 feet. Eelgrass on the Pacific coast is of two varieties. *Zostera marina* var. *marina* has leaves up to 4 feet in length and grows in shallow water near the extreme low-tide level. *Z. marina* var. *latifolia* has leaves up to 12 feet or more in length and grows in water up to 20 feet in depth. In the 1930s, a parasitic infection practically eliminated eelgrass along the Atlantic coast. As a result, the scallop population died and the brant, which fed mainly on eelgrass, almost became extinct. Eelgrass is now regaining its former abundance. The black brant of the Pacific coast also feed on eelgrass in the estuaries of Oregon and California on their winter migrations from Arctic breeding grounds.

Other "sea grasses" are abundant in the green water world of estuaries and shoal water along the Atlantic and Gulf coasts. Turtle grass (*Thalassia testudinum*) is common in semitropical waters. It belongs to the frogbit family and is an aquatic herb distinguished from other members of the family by its broad strap-shaped leaves, ⅛ to ¼ inches wide. Many of us know one of its relatives, waterweed (*Elodea canadensis*), with its many whorled leaves, grown in aquariums with tropical fish.

Turtle grass takes its name from the green turtles that eat it. Its great underwater forests are home for whelks and other snails, starfish, blue crabs, sea horses, and pipefish, which in the vertical position resemble the leaves of turtle grass. Nearly a hundred species of smaller seaweeds have been found attached to Florida turtle grass.

Widgeon grass (*Ruppia maritima*), a common marine flowering plant, is considered to be one of the most valuable of all submerged species. Fish use it for food and shelter from prey species. The numerous flowers and seeds that it produces are also eaten by marsh and shore birds. Widgeon grass produces leaves directly from the rhizome instead of from stems. The flat leaves are only 1/25 to 1/16 inches wide.

Manatee grass (*Cymodocea filiforme*) grows along the Gulf coast from Florida to Texas in shallow-water bays. Its long round leaves superficially resemble onion leaves. The plant remains submerged and reproduces mainly by vegetative growth. Shoal grass (*Diplanthera wrightii*), which grows in shallower water, has narrow flat leaves 1.5 to 2 millimeters wide and roots that often terminate in fleshy, starchy, tuberlike swellings. These two sea grasses are valuable sources of food for wildlife along the seashore.

In contrast with the sea grasses that flourish in the relatively gentle waters of bays and sounds, two species of surfgrass endure the full force of the waves on Pacific shores. *Phyllospadix scouleri* has flat leaves less that 3 feet long, while *P. torreyi* has oval or round leaves that are 10 feet long. These plants grow on rocks of the subtidal zone. Their flowers are borne in catkinlike clusters, and pollination occurs under water.

On Atlantic beaches the land vegetation begins near the highest wave action. Since the plants are annuals, it matters little if their dead stems and roots are washed away by winter storms. Their seeds can start a new generation when summer comes again. Annuals are adapted in various ways to life on the beach. Some are ground-hugging plants with exceedingly short stems, such as the seaside spurge, *Euphorbia polygonifolia,* mentioned previously. Whenever I see it I am reminded of the pesky purslane (*Portulaca oleracea*) that grows so persistently in my vegetable garden. Both are

Large algae such as *Macrocystis, Postelsia, Nereocystis,* and *Laminaria* are brought to shore by storm waves. Some of these have large leaflike blades, some have long ropelike stipes, while others are shaped like palm trees. Frequently the holdfast is so strongly attached that it brings its rock to shore along with the plant.

succulent, freely branching, profusely seeding, and sun-loving plants. If my purslanes are covered with straw, they die. If the spurge is covered with sea wrack, it, too, dies.

Since tide-edge plants usually are widely spaced, they do not compete with one another. Their main tasks for survival are to endure the salt spray, resist excavation by moving sand, and retain moisture in their succulent leaves or slow evaporation by the hairy coverings on their leaves and stems. The sea rocket (*Cakile edentula*), for example, tends to be a solitary plant. It is easy to recognize since it belongs to the mustard family and has bluish or lavender flowers, short two-jointed seedpods, and fleshy, slightly toothed leaves. It grows all along the sandy Atlantic shores. You should also look for it if you tramp along the shores of the Great Lakes.

The green fringe at the upper beach includes many common and conspicuous plants. Among these are ragweed, curled dock, and mullein. The seaside goldenrod (*Solidago sempervirens*) is a robust plant with thick leaves and golden yellow flowers from August until frost. The last time I saw them, in early May, their stiff dried leaves and stems stood erect on the sand berm above the beach at the Maryland–Delaware line. Their sturdy roots had held them in place through all the winter storms.

Where the sands have become somewhat stabilized on the leeward sides of foremost dunes you should look for colonies of beach peas (*Lathyrus japonicus* var. *glaber*). The violet-purple flowers grow in clusters, and the blossoms are present throughout the summer. The beach pea spreads by underground stems. On Cape Cod, the salt spray rose (*Rosa rugosa*) also grows on the foremost dunes. The leaf margins are recurved, the petals are purplish rose to white, and the shrubby plant produces red fruits, or hips, which are conspicuous in autumn.

Along with the roses and beach peas you may find the dusty miller (*Artemisia stelleriana*) with its showy spikes of yellow flowers. The pale green leaves have a powdery appearance owing to their covering of tiny white hairs. A relative of the dusty miller is wormwood (*Artemisia caudata*), with linear divisions of the leaves and racemes of small greenish-yellow flowers. The root is biennial, and the flower stalks arise from it in the second year. Wormwood is a sand-loving species and occurs inland as well as along the sea coast.

Behind the beach on sandy shores the dunes arise tier on tier. Here, the sand-binding grasses appear. More will be said about these in Chapters 6 and 8. The dunes are frequently colonized by some of the hardy plants that grow along the beach. Among these are the seaside goldenrod, beach pea, dusty miller, and the salt spray rose. On the stabilized dunes are beach plums, poison ivy, golden aster (*Chrysopsis falcata*), and beach heath (*Hudsonia tomentosa*). The latter is a low shrubby plant covered with downy hairs and leaves $\frac{1}{16}$ inch long, pressed together like shingles on a roof. Still farther inland are pines and oaks.

On New England's rocky shores the forests come down to the water's edge. On the relatively few areas where pitch pine (*Pinus rigida*) produces humus on stabilized dunes, the trees provide shade for huckleberry, blueberry, hazelnut, bearberry, and sweet fern (*Myrica asplenifolia*). Sweet fern is not a fern; it is a shrub with scented linear leaves, globular female catkins, and heart-shaped sterile catkins on the same plant or sometimes on separate plants. It beongs to the sweet gale family. Its relative, bayberry (*M. pennsylvanica*), also has fragrant leaves and wax on its berries. The wax is used in making bayberry candles. This shrub is common to Cape Cod.

From Virginia southward and across the Gulf of Mexico from Florida to Texas, a variety of plants excites the admiration of the seashore wanderer. They are too numerous to be listed anywhere but in the botany books, but some are so conspicuous, abundant, or characteristic of local habitats that one learns to associate them with marshes, stream banks, chênières, dunes, and sandy beaches.

The beach morning glory (*Ipomoea stolonifera*) is easy to recognize by its trailing stems and white flowers. Look for it above the high-tide line, where it is confined to sandy beaches from Mississippi to Texas. Here, too, you will find Drummond's evening primrose (*Oenothera drummondii*), a densely pubescent trailing herb with bright yellow flowers 2 inches wide. It also grows on the chênières and in the prairies of southwestern Louisiana. Another chênière inhabitant is narrowleaf gromwell (*Lithospermum incisum*), a perennial herb 12 inches tall with long narrow leaves inrolled at their margins. The pale yellow corolla tube is 1 inch long and ends in five lobes finely shredded at their tips. It is a member of the borage family.

Along with the seaside goldenrod you may see the lazy daisy (*Aphanostephus skirrobasis*), a much-branched annual with pubescent stems and leaves and numerous flower heads with white rays and yellow disks. It grows about 12 inches in height on sandy beaches. Here, also, and in salt marshes from Mississippi to Texas, the salt pennywort (*Hydroctyle bonariensis*) produces round peltate leaves on stalks that rise from creeping stolons. The inflorescence is a much branched umbel—the plant belongs to the parsley family—with many tiny flowers, scarcely ¹⁄₁₆ inch in diameter.

The salt matrimony vine (*Lycium carolinianum*) grows on slightly elevated marshland. It is a woody shrub with purple flowers which give rise to fleshy red berries ½ inch in diameter. In the thousands of marshlands where these wildflowers occur, the dominant vegetation consists of grasses, sedges, and rushes.

The Pacific coast has so many communities and ecosystems along the shore that it is a flower watcher's delight. There is marked zonation of vegetation above the sea. There are miles of sandy shores, towering dunes, fresh- and saltwater marshes, and ocean-facing bluffs where ferns, shrubs, and showy wildflowers flourish according to their habitat requirements.

On the coastal strand from southern California to British Columbia, the sand verbenas are made conspicuous by their expanded calyxlike bracts surrounding the flowers, which grow in heads. *Abronia latifolia* has yellow petallike parts. *A. maritima* flowers are

In grassy meadows on terraces above the sea such colorful plants as butterweed groundsel, various clovers, paintbrushes, and asters make extensive flower gardens on Pacific shores from California to British Columbia.

red to purple, and the fleshy leaves are covered with sticky hairs. Other showy species are the wallflower (*Erysimum franciscanum*), with yellow to cream-colored flowers, the locoweed (*Astragalus pycnostachyus*), and the senecio (*Senecio bolanderi*). Throughout the West there are many species of *Abronia* and *Astragalus*, but the two mentioned here are the ones frequently seen along the coast from southern Oregon to central California.

Two plants the coastal traveler cannot fail to see are the Hottentot fig (*Mesembryanthemum edule*) and the sea fig (*M. chilense*). The latter produces rose-colored flowers 2 inches in diameter, while the former produces yellow flowers up to 3 or more inches in diameter. Both have fleshy leaves. Since they grow nicely on sandy soil they are often planted on road banks to control erosion. A third species, the ice plant (*M. crystallinum*), produces whitish flowers less than 1 inch across. It is an annual with broad succulent leaves covered with minute colorless nipple-shaped projections. Look for it on the sea beaches.

For seashore explorers who wander away from the beach into the dunes or up on the slopes to the benches and grasslands, the variety of wildflowers, sedges, grasses, and shrubs is almost endless. Only a few examples can be given here. But local manuals, such as *Shore Wildflowers of California, Oregon and Washington* by Philip A. Munz, can introduce you to dozens of easily recognized seashore plants.

In the flat marshy areas between the Oregon dunes, for example, you may find golden-eyed grass (*Sisyrinchium californicum*), with its broad basal leaves with parallel veins and bright yellow flowers. It is not a grass but a member of the iris family. In the same moist areas do not be surprised if you find a native orchid, hooded ladies' tresses (*Spiranthes romanzoffiana*), growing within the sound of the booming surf. Here also you may find the small creeping buttercup (*Ranunculus flammula* var. *ovalis*) with its small yellow flowers.

On sandy areas and on hillsides above the ocean are various members of the pea family. The beach pea (*Lathyrus japonicus* var. *glaber*), with bright green leaves, and the gray beach pea (*L. littoralis*), with white silky leaves, range from California northward along the Pacific shore. The seashore lupine (*Lupinus littoralis*) can be recognized by its stems that creep over the sand, its silky leaflets, and its purplish-blue flowers. The giant vetch (*Vicia gigantea*) with its sixteen to thirty leaflets and yellowish flowers creeps on the sand or climbs over logs and driftwood washed up by the winter storms. Another legume you cannot miss is Scotch broom (*Cytisus scoparius*), a shrub that grows to 6 feet in height and presents a blaze of yellow when its flowers are in bloom. It has been naturalized from Europe and is used to stabilize dunes. Scotch broom is now established along the shoreline from Washington to California and inland to such cities as Portland, Oregon.

If you ever drive over Neahakahine Mountain, north of Tillamook on Highway 101 along the Oregon coast, look for the spectacularly beautiful foxglove (*Digitalis purpurea*) that grows on the hillsides. This plant is a member of the figwort family which includes the paintbrushes, snapdragons, monkey flowers, and penstemons. These plants also live along the coast.

The foxglove reaches a height of 6 to 9 feet and has soft hairy leaves which form a large basal tuft in the first year. The inflorescence appears in the second year in a terminal raceme. The corolla is somewhat two-lipped, inflated on the lower side, purple, with the lower bottom of the throat whitish and purple dotted. It is one of the prettiest flowers I know. Foxglove is naturalized from Europe and now is established near the coast from California to British Columbia. It is the source of the digitalis glycosides that are used in treating congestive heart failure and other diseases of the heart.

Goat's beard (*Aruncus vulgaris*) grows in moist woods, logged-off lands, and along the coast. I like its other common name, seafoam, which seems appropriate when this plant grows down to the water's edge as it does along the south shore of Vancouver Island, British Columbia. It is a perennial herb of the rose family, possessing coarsely toothed leaflets that resemble those on our garden roses. Its cream-colored flowers, however, are very small and numerous on long slender racemes that rise 3 to 6 feet above the leafy stems. Look for it from California to Alaska.

Mangroves are seed plants that capture the sea. These trees have unique structures and reproductive capabilities that enable them to endure the salinity of the ocean, to extend coasts, and to build islands. They protect the seashore from excessive erosion and furnish habitat for wildlife that ranges from shrimps to fish to exotic birds.

Three kinds of mangroves grow along our southern shores. The black mangrove (*Avicennia nitida*) occurs along the Florida coast north to St. Johns and Levy counties and along the coast to southern

Louisiana and west to southern Texas. The red mangrove (*Rhizophora mangle*) is common along the coasts of central and southern Florida and in the Keys. It also grows along the Pacific coast from Lower California to Mexico, south through Central America, and in Bermuda and the West Indies. The white mangrove (*Laguncularia racemosa*) grows along the coasts of central and southern Florida, including the Keys, as well as along the coasts of the West Indies, South America, and Africa. These three species of trees, so tolerant of salt water, are not closely related. They belong to different plant families.

The red mangrove is the bizarre one of the group. It grows the farthest into the sea and maintains its position against the force of the waves by means of extensive rhizophores or prop roots. These form such extensive tangles, aptly called "monkey puzzles," that they actually slow tidal currents and promote the depositing of mud to the extent that land rises beneath the trees. The seeds sprout

Mangrove trees produce a veritable "monkey puzzle" of roots that can resist even hurricane waves. Many islands in the Gulf area owe their origin to mangrove trees, which actually invade the shallow waters of the sea.

while still attached to the tree. The hypocotyl, or root portion of the seedling, may reach a length of 2 feet before the seed drops into the mud or floats until it touches shallow bottom. Then it quickly establishes new roots and adds to the vast tangles that have built land for centuries along the southern coasts.

The white mangrove sometimes grows in company with the red mangrove as it invades the sea. More frequently it grows nearer to dry land with the black mangrove. In this seaward community, various marine plants associate with the trees. Among these are turtle grass and manatee grass.

The black mangrove forms a zone nearer the shore and follows the red mangrove as it forms firm soil. It grows best from slightly below high-tide level to the soil line above the sea. Here, certain saltmarsh plants, such as the glasswort (*Salicornia virginica*) and smooth cordgrass (*Spartina alterniflora*), also find favorable growing conditions. Even farther toward the land is buttonwood (*Conocarpus erectus*), which is not a mangrove but a woody plant that forms a transition between the mangroves and the drier coastland vegetation. The cabbage palm (*Sabal palmetto*) and the strangler fig (*Ficus aurea*) commonly grow in the buttonwood community.

In this zonation, or succession, from sea to land, the red mangrove begins to die off as the muddy substratum builds up, the water becomes shallow, and marshy conditions prevail. It is believed that lack of oxygen in the mud contributes to the die-off. The black mangrove then invades and grows successfully because of its specialized root system, which develops pneumatophores. These structures are upright projections of the roots. They contain air channels that allow oxygen to pass from the air to the submerged roots. Ultimately, the mangroves are replaced by tropical forest or marshland. In Florida, the succession may even proceed to sand dune communities or to pinelands.

The mangrove forests provide fabulous habitats for wildlife. Their fallen leaves settle to the ocean bottom, where they are broken down by anaerobic bacteria into basic nutrients for microorganisms, which in turn provide food for the larvae of shrimp, crabs, oysters, and barnacles. Oysters, barnacles, mussels, and other sedentary species attach themselves to the roots and stems of the red mangrove. Young fish find food and protection in the intricate tangle, as do the more mobile hermit crabs, fiddler crabs, and larger fish that come in from the sea. In the mangrove jungles of the Ten Thousand

Islands up the west coast of Florida to Cape Romano, the brown and pink shrimp first grow to maturity before they return to the sea to be taken by fishermen for the American market.

The mangrove canopy is home for epiphytes, reptiles, mammals, birds, and insects. Mosquitoes are legion. Cattle egrets and boat-tailed grackles roost in the branches. Roseate spoonbills and herons nest in the jungle. Black-necked stilts and other shorebirds feed on the tidal flats among the mangrove seedlings. The mangroves are rookeries for thousands of other tropical and semitropical birds. They are nurseries for a multimillion dollar fishery. And they are a protection from erosion by the sea. But in unprotected areas the exploiters are ripping them out of the ground for housing projects, boat landings, and tourist hotels.

If you sit quietly someday at the edge of the shore and observe people who explore beaches you will see that they tend to concentrate their attention on two areas, the sea and the strand or rocks beneath their feet. They seldom stray far into the dunes, climb the long slopes above the tide marks, or explore the forested shores. Many people miss seeing the beauty of trees, the mystery of bogs, or the richness of the land flora that exists in peaceful solitude while they are still within hearing distance of the murmuring waves or the thundering surf. They are seemingly unaware that you still can find a wilderness tucked between the megalopolis and the sea where strange, beautiful, and magnificent plants adorn pleasant environments.

Walk away from the seashore in Maine and see what lies just beyond the shore. Leave the ocean's edge with its rocks and sand where the saltwater glasswort and sea blite grow in little clusters. Pass by the orach with its arrowhead-shaped leaves on red-streaked stems. Climb over the crustose lichens on rocks above the tide line and meet the bayberry 50 feet from the water. Observe its tangled shrubby branches with pale gray berries. Continue onward to the small junipers and raspberries that fringe the spruce forest. The goldenrod will be there at the edge of the spruce grove.

Enter the forest and enjoy the softness of the needle-strewn floor. Where sunlight enters, look for carpets of haircap moss or beds of lady ferns or cinnamon ferns. Don't miss seeing the parmelia lichens. And in this paradise for mushrooms, look down for waxy clitocybe (*Laccaria laccata*) and the fluted amanita poking up through the needle carpet. And look up at the old-man's-beard on dead spruce branches.

Remember that these small plants are not insignificant in the realm of nature. The Northeast coast, at the end of the last glaciation, was once a lifeless landscape of rocks and troughs gouged in the land. Later, it was waterlogged from Connecticut to Newfoundland. Then the lichens, mosses, and sedges created a substratum for herbs, shrubs, and sun-loving trees, such as oaks and maples. Then came the hemlocks and spruces that could reproduce and grow in their own shade. Even in poor soil the spruces and balsams could gain a foothold and maintain their dominance near the sea. When fire destroyed the forest, it was followed by blueberry bushes, birch, sumac, poplar, and wildflowers and ultimately by spruce and balsam.

An East Coast semiwilderness, between the miasma of megalopolis and the sea, is the New Jersey Pine Barren country. Here, more than a half million acres of pitch pine and blackjack oak, dwarfed by the fires of centuries, shelter a menagerie of plants, mammals, fish, and picturesque outdoorsmen who work the country for cranberries, cattail decorations, sphagnum moss, and an occasional duck. The towns within the barrens are more or less self-contained and the roads are poor. The soil is little more than sand, especially where the pines come down to the tidal marshes.

The naturalist and botanist will find this wilderness worth exploring to see the cranberry bogs and maybe to catch pickerel, bass, and catfish. Here, the beavers still build dams, and deer, mink, raccoons, and gray foxes make their homes. The herpetologist can find rattlesnakes, pine snakes, milk snakes, and corn snakes. And for the botanist, *Arethusa* is only one of twenty orchids that grow in the pinewoods and bogs. If you can, explore this land soon, for exploiters and their bulldozers are beginning to rip away the trees and the sand. And engineers want to build a road through the Pine Barrens from New England to the South.

The islands along the East Coast exhibit many familiar plant friends for visitors who wander away from the immediate seashore. On eastern Long Island and Cape Cod, salt spray roses (*Rosa rugosa*) intermingle with the beach plums and bayberries. Beach peas and seaside goldenrod are not confined to the shore zone, and clumps of *Hudsonia* or beach heather and mats of bearberry carpet the sand in the hollows. The flora also includes various mushrooms and showy seasonal herbs.

On Hatteras Island, inland from the shore, you will see a savanna with clumps of dusty-olive wax myrtle, yaupon or sea island holly, and groundsel bushes. Yaupon (*Ilex vomitoria*) is a low tree with small, elliptical, vaguely scalloped bright green leaves that grow on whitish-green branches. The shining scarlet berries remain on the female plants all winter. The leaves were the source of the "black drink" of the coastal Indians, who dried them in vessels over slow fires and then steeped them for a drink for ritual ceremonies.

Southward from North Carolina to southern Florida and westward along the Gulf of Mexico, the landward-wandering botanist will encounter a bewildering variety of plants in the live-oak borders, the sawgrass margins, and the open grasslands and groves of broadleaf evergreen trees. The groundsel tree (*Baccharis halimifolia*), which belongs in the composite family, grows to a height of 3 to 9 feet. It is conspicuous in autumn, when its white pappus extends from the seed heads as the pyramidal panicles mature.

The wax myrtle (*Myrica cerifera*), with its bony nuts encrusted with white wax, may be seen on sandy lands from Maryland to Florida to Texas. It is a relative of sweet gale (*M. gale*), which grows on swamp borders from Labrador to New England and in the Great Lakes country. On sandy lands of the Gulf region you will see palmettos, yuccas, and many herbaceous plants with flowers of surpassing beauty.

Now to the Far West. On the southern coast of California the chaparral is the predominant plant type. Originally, this mixture of shrubby species, also called sclerophyll bush vegetation, extended from Baja California to northern San Luis Obispo County, California. The representatives of many plant families in this 50- to 60-mile-wide coastal strip are characterized by drought resistance and susceptibility to fire. Most are evergreen with thick leathery shining leaves. Only a few are deciduous plants.

Chamise, a member of the rose family, is the most widespread of the chaparral shrubs. Its pale green leaves are short and needlelike and its large, showy white flowers bloom from February to July. A dozen species answer to the name of another common plant, manzanita, in the chaparral country. These shrubs of the heather family sometimes approach the size of a small tree. As a group they are characterized by smooth, dark, red-brown bark. In spring, they are covered by small pink or white flowers.

The California scrub oak, an evergreen with short prickly leaves, also grows in the chaparral zone. Other shrubs that add variety are buckbrush (*Ceanothus cuneatus*), birchleaf mountain mahogany (*Cercocarpus betuloides*), and bearberry (*Arctostaphylos uva-ursi*). Lupines with pealike flowers colored magenta, violet, or blue, or blue and white, grow near the rocky cliffs. The yellow tree lupine (*Lupinus arboreus*), with ten or more leaflets radiating from the leaf petioles, bursts into bloom in April on flats and hills above the sea. And near the coast, in moist soils, the coast figwort (*Scrophularia californica*) thrusts its square reddish stems with triangular saw-toothed leaves to heights of 5 feet or more. The figwort has bell-shaped red to maroon flowers with protruding yellow anthers, and you can see it from Los Angeles to British Columbia.

The giant horsetail grows on rocky cliffs, in moist sand, and in burned forest areas along the Pacific coast. Young plants arise from running rootstocks and, when fertile, are terminated by spikelike fruiting tips that bear spores. The mostly hollow stems have sheaths at the joints. The horsetails are close relatives of the club moss family.

From northern California to Alaska, big trees come down to the seaside in forests that have no equal among forests of the nation. The southernmost trees are the gigantic coast redwoods (*Sequoia sempervirens*). The largest redwoods have trunk diameters of 10 to 15 feet and thrust their heads skyward 200 feet or more. In the damp shade and cathedral silence beneath their crowns, luxuriant ferns and thick underwood grow in profusion. Mosses and ferns grow on giant fallen logs that may remain for a century or more before they rot and return their substance to the earth.

Away from the sea, the redwood forest has an understory of lesser trees. Where sufficient light filters through the high canopy, red alder, wax myrtle, tan oak, and chinquapin grow in the cool twilight of the forest. In more open spaces, huckleberries and showy rhododendrons produce thickets. Some of these shrubs, on the seaward edge of the forest, extend their growth out to the edge of the sandy shoreline.

Northward, on the coast of Oregon and Washington, another forest of mighty trees borders the edge of the continent. Here, the mammoth Sitka spruce (*Picea sitchensis*) grows in company with western red cedar (*Thuja plicata*), and western hemlock (*Tsuga heterophylla*). Other trees that grow in this giant forest, but mainly in the transition region from fogbound coast to higher mountain slopes, are Douglas fir (*Pseudotsuga menziesii*) and red alder (*Alnus rubra*).

In the Puget Sound country, where the giant trees come down to the shore, the richness of flora creates a paradise for plant lovers. In the rain forest of almost perpetual gloom, the trees may reach ages of 1,000 years. Trees 100 years old grow in rows on fallen logs, which eventually disintegrate and leave their progeny standing on great arches of roots, beneath which one can walk with ease.

On the forest floor are ferns and mosses that form spongy carpets deeper than any shag rug ever made by man. And on the trunks and branches of trees are ferns, lichens, and innumerable species of mosses that grow as epiphytes, "plants upon plants." These nonparasitic species derive their nourishment from the dissolved minerals contained in the water that drips down from the canopy, or is brought in by drenching rain from the sea. The winds come from the sea, but on the forest floor even the force of mighty storms is broken. In periods of calm, one can stand in the pale gloom and feel that time here is without end.

5

The Strange Lives of Sea Creatures

THE GREAT PROBLEMS ALL LIVING CREATURES FACE are to obtain food, protect themselves from predators, and produce young so the species can persist. One of the wonders of nature is the variety of ways in which seashore animals solve these problems. Food is of paramount importance. Without food, no plant or animal can maintain its existence or procreate. When we consider the many kinds of animals along the seashore, each with its own specialized adaptations for survival, the ways of life become almost incomprehensible.

Some animals and plants are adapted for floating and thus are at the whim of wind and sea. Burrowing animals find food and protection in the mud and sand of shores and estuaries. Sessile animals attach themselves to rocks and remain without moving throughout their lifetimes. Swimming animals, such as fish, squid, and even certain clams, are free to move to other environments when they need new food supplies or special habitats for their young. The creeping animals, including snails, sea hares, sea stars, urchins, and worms, also have evolved in different ways to obtain food and to reproduce their kind. There is space here to discuss the strange lives of only a few sea creatures.

Few of us ever see or have the energy to dig up and examine some of the weird sausage-shaped worms that can be found at the shore. These worms dwell in U-shaped mud burrows with two openings that extend to the surface of the bay bottom. They get oxygen from water, which they pump through their tunnels by rhythmic contractions of their bodies. Some have a proboscis that can extend to the surface of the tunnel to pick up detritus for food.

The sea snails, like our garden snails and slugs, eat with a radula, or filelike rasping organ, which scrapes algae from rocks, or munch the fronds of marine plants. Carnivorous snails have strong teeth that are capable of drilling holes through the hard shells of oysters and clams. They then thrust their proboscis through the hole into their victim's body—sometimes poisoning it with venom and thus softening the flesh, so it can be drawn out more easily.

Even in a tiny area of the seashore you can observe a variety of methods that animals use to obtain food. On a sandy beach, the clam strains out microscopic forms, unaware that the whelk will drill into its shell to obtain its meat. But the whelk is unaware that the blue crab may crack its shell with powerful claws and consume its succulent muscles and internal organs. And, finally, the crab is unaware that man may catch it and sell it for food. This is the story of life at the seashore and in the sea itself. Eat and be eaten.

The crabs manage to persist, however, in spite of the fish, gulls, and even octopuses which are their enemies. There are many kinds of crabs and each lives its own strange life, depending on its need for food and shelter. You can find them in tide pools, in cliff grottoes, and among the rocks at low tide.

One of the pleasures of learning about the lives of sea creatures is talking with people who have lived near the ocean for many years. One April day, when spring had arrived early on the coast of Georgia, I met a gentle old black man who had been fishing. He seemed to be enjoying the melody of birds singing in the live oaks. As we talked, his words revealed a lifetime of observing wild creatures that inhabited the waters and the beaches of the "golden seashore."

"The dogfish are bitin' good in Bull Bay," he said. "And the fiddlers are fiddlin' in the mud flats." The fiddlers, of course, were the fiddler crabs. "Come summer," he continued, "the hollyhock flowers in the marshes will be white with pink eyes." He meant swamp mallows (*Hibiscus aculeatus*).

He admired the Spanish daggers (*Yucca aloifolia*) that grow on the dunes and hammocks and produce clusters of drooping white flowers above their bayonet-shaped leaves. He told me there was good fishing where the alligator bonnet (*Nymphaea odorata*), the water lily of the coastal ponds, produced its large submerged leaves in deep water while the white or pinkish flowers floated on the surface throughout the summer. "Then the glass plants along the salt creeks will be hidin' the fiddlers," he added. These were the perennial glassworts (*Salicornia virginica*), which produce semiwoody stems.

We found the fiddler crabs, which live along the coast from Cape Cod to Florida. These crabs excavate burrows by carrying mud balls to the surface in salt marshes or on mud flats. The western species (*Uca crenulata*) along California shores digs to depths of three or four feet. At least three species live along the Atlantic coast. Non-breeding males and females retire to temporary burrows during high tide. At low tide, thousands emerge to feed on organic matter in the mud, at which time they can be herded like livestock.

Female fiddler crabs mate once each lunar month before one of the semimonthly spring tides. A study of the fiddler crab *Uca pugilator* on the west coast of Florida by John H. Christy suggests that this is an adaptation whereby the final larval stage will settle on substrates suitable for adults. This is one of the biological rhythms exhibited by so many animals of the seashore.

There are many kinds of crabs. Hermit crabs, so common in tide pools, live in snail shells. They are constantly faced with the problem of finding larger shells as growth forces them out of their old homes. The sand crab or mole crab of the open sandy beaches is a burrowing expert.

Practically everyone who has ordered oysters in a restaurant along the Atlantic coast has seen the little pea crab, which lives in burrows of sea worms and in clams and oysters. Usually, only one of these little crabs (*Pinnotheres ostreum*) is found in an oyster at any one time. They turn pink, as do other crabs, when they are accidentally steamed or cooked with oyster soup.

The ghost crab (*Ocypode albicans*), the speedster of the crab clan described in Chapter 2, is a denizen of the high beaches above the high-tide mark from New Jersey to the Gulf of Mexico. Its color matches that of the sand where it lives and makes its burrows.

There are many other kinds of crabs, but two are famous because they are among the most delicious of seafood items. One is the Dungeness crab (*Cancer magister*) of Oregon and Washington, and the other is the Atlantic blue crab (*Callinectes sapidus*). The fascinating life histories of these two crabs are similar. That of the blue crab and the crabbing industry is recorded in appealing detail by William W. Warner in his Pulitzer Prize-winning book, *Beautiful Swimmers: Watermen, Crabs and the Chesapeake Bay.*

The courtship of the blue crab is a prolonged exercise in lovemaking rarely exceeded elsewhere in the animal kingdom. A female crab molts eighteen to twenty times, and the job of getting out of her old hardened shell is somewhat of a gymnastic feat. Her last molt brings her to sexual maturity. When this time comes, the male displays himself on his walking legs and stirs up sand with his swimming legs. Then he carries her for two days or more in a cradle made by his legs. He releases her while she again molts and strengthens her muscles, after which they embrace face to face for copulation. She is again cradled and carried for at least two more days while her shell hardens. When released, she swims to the winter resting place, where she buries herself in the muddy or sandy bottom.

The sperm packet of the male is not used until spring. The two million fertilized eggs from each female hatch into microscopic larvae that become part of the plankton—prey for other minute predators as well as a multitude of animals that filter food from the sea. These almost invisible specks of life are called zoeae.

The zoeae go through seven or eight larval molts as they develop hairs and various appendages. After the last larval molt, a metamorphosis changes the tiny animal into a megalops with stalked eyes, walking legs, and a shrimplike tail. With its first molt, the megalops becomes a crab and begins its life of molting and growing larger until it too can help reproduce its kind.

Among the strangest creatures of the sea is the octopus. In shape and structure it is one of the most spectacular invertebrates in the marine environment. Unlike other animals without backbones, the octopus has a brain and sense organs comparable to those of many vertebrate animals that swim, walk, or fly. It can learn from experience; it can remember; and although it cannot hear, it has excellent vision and color sensitivity. And its sense of touch is so keen it can distinguish textures of rocks and other materials around its home.

The octopus is a member of the phylum Cephalopoda, which includes the squids and the chambered nautilus. The latter has a shell. The squids, with their ten arms, have a soft internal remnant of a shell. The octopus has none. But it has eight arms with hundreds of suckers, a well-muscled mantle cavity containing gills, a head and eyes, and a pair of jaws shaped like a parrot's beak. Its mouth also is equipped with a radula, with which it can rasp through clam and abalone shells and exude a poison that anesthetizes its prey. Its body has no fins.

The octopus's saclike body has a siphon opening near the head. When it takes water into the mantle cavity, it can trail its arms behind the bulbous part of its body, squirt the water through its siphon, and proceed by jet propulsion. It also can swim slowly by waving the membranes or webs between the arms near their point of attachment.

If an enemy seems to be near, the octopus expells a cloud of ink, which may serve as a smokescreen or give the predator the impression that the dark cloud is the octopus itself. The ink can also dull the sense organs of deadly enemies, such as the moray eel, whose principal food is octopuses.

Not the least of the accomplishments of the octopus is its ability to change color. Its "skin" contains many chromatophores, or pigment patches, which are operated by muscles. When alarmed, protuberances develop on the body and its color may rapidly change through a series of blacks, reds, yellows, and browns. Even baby octopuses are capable of color changes.

The octopus is a retiring animal and is not inclined to attack

Fiddler crabs dig holes in wet, sandy soil and mud by excavating small pellets. They disappear during high tides and swarm over the mud flats at low tide. The large claws attract other crabs during mating displays.

human beings. Biologists routinely handle them without getting bitten; however, if handled carelessly, they can bite and the bite can be serious. The blue-ringed octopus, a small species that lives along the coast of Australia, so named because it displays blue rings on its skin when disturbed, possesses a neurotoxin known to be fatal to humans. The poisonous substance, maculotoxin, aids the octopus in killing the small crayfish and crabs on which it preys. This toxin may be as lethal as a certain chemical my old chemistry professor used to warn us about. "It is so potent," he said, "a drop on the tongue of a dog will kill a man."

The octopus's food consists mainly of scallops, mussels, and clams. It will also eat fish if it can capture them. When caught, the prey is covered with the web between the bases of the arms and is paralyzed by poison secreted from the octopus's mouth. The octopus's parrotlike jaws are strong enough to kill small animals, such as crabs.

At breeding time, one arm of the male enlarges and becomes modified so it can be charged with a packet of spermatozoa. These are inserted under the mantle of the female. The female later lays eggs in clusters, which are attached to a rock. The female guards the eggs until they hatch; according to some reports, she then dies. The young octopuses do not undergo a metamorphosis, as do crabs and many other marine animals. They appear as miniature adults when they leave the egg and even at that tender age can squirt ink.

A recent discovery by Dr. Jerome Wodinsky, a Brandeis University psychologist, indicates that the glands between the eyes of the octopus function like the pituitary glands of most land animals in the control of hormone production, including those involving sex and reproduction. When these "optic" glands are removed, female octopuses do not die immediately, but double their life span and weight. This intriguing finding suggests further study of the aging process in animals.

Because of its complex brain, with thirty distinct lobes, and brain cells so numerous they compare with higher vertebrate animals, the octopus has been the subject of many experiments and anatomical studies. Octopuses can learn rapidly. They respond to reward or punishment. Their memories are long-lasting. And their reflex activities are highly organized, especially in their arms, which have hundreds of sucker disks. Truly, the octopus is a special invertebrate.

Some of the large octopuses are denizens of the deep oceans. But several species live along the Atlantic coast and in Florida waters. The mud-flat octopus (*Octopus bimacubides*) is a small species found southward from Santa Barbara, California. It is common under rocks in the outer tidelands and can be found occasionally in crevices in tide pools. Octopus hunters, who apparently relish this kind of seafood, capture them by pouring chloride of lime, laundry bleach, or copper sulfate into the hiding place. When the octopus emerges it is caught by hand or with a hook on a wooden stick. The use of chemicals to capture fish is illegal, but they have been used so often that the octopus population has been depleted in some localities. Occasionally, octopus limbs are on sale at the meat markets in chain stores.

The largest species of these remarkable creatures is the giant Pacific octopus (*Octopus dofleini*), which may have an arm spread of 15 feet and weigh 100 pounds. It is especially common in Puget Sound and may be found all the way from Oregon to Alaska.

The associations between different species that live in this world can be incredible. We do not have to go to the sea to find them. Many are right here at home. Consider the some 60 million dogs and cats that people keep in the United States. Some of these relationships involve bonds of affection. Others involve protection of property by watchdogs. The possession of horses provides pleasure for riders and an opportunity to gamble on races—an attribute apparently limited to humans alone.

Man has many other relationships with animals. Rats live in his garbage dumps. Mosquitoes infest him with parasites that produce sleeping sickness or malaria. The beef he eats is made possible by bacteria and ciliate protozoans that digest cellulose in the rumens of cattle, permitting the cattle to grow, live, and reproduce. Even the flowers in man's garden are affected by aphids that have been nurtured through the winter and placed on plants by ants.

Looking at the sea, with its thousands of species, each with its niche to fill, it is little wonder that natural selection through millions of years has evolved a variety of relationships among animals, ranging from harmful predation and parasitism to commensalism or mutualism, where one or both parties contribute food, protection, or shelter in a beneficial relationship.

Scientists have studied animal relationships for years and have developed terms to describe them, many of which are bizarre, complex, and beyond explanation. The word commensalism refers to animals living together, sharing a source of food, and doing no harm to each other. This is illustrated by the small fish (*Nomeus gronovii*) which lives among the tentacles and stinging cells of the Portuguese man-of-war (*Physalia*). The small fish apparently is immune to the poison of the man-of-war; other fish that attempt to eat *Nomeus* are stung and captured by the protective tentacles and serve as food for both creatures.

This type of mutually beneficial partnership is also described as symbiosis. Symbiotic relationships include cooperation between plants and animals as well as cooperation between animals of different species. When both benefit from the association it is called mutualism. The lichen, for example, is a combination of alga and fungus in which the chlorophyll of the alga supplies food and the body of

Sea anemones are common in tide pools and tidal channels on rocky shores, where they become embedded in sand. When the tentacles are open, the animal presents the appearance of a beautiful flower. The green in the tentacles results from the presence of one-celled algae in the tissues of the anemone. When the tide is out, the anemone folds in upon itself and all its beauty disappears.

the plant provides moisture, attachment to the substratum, and support for the joint structure of the two plants. In the sea, certain small chlorophyll-bearing animals rise to the water's surface during the day to receive the sun's energy and sink to the depths during the night. The animals apparently provide the motive power for these changes in position in the sea.

Parasitism exists when one animal takes advantage of its partner in such a way that its host is harmed or even killed. Parasites, of course, would defeat their own purpose in life if they killed all the animals they attack. This is prevented to some extent by a form of mutualism whereby the parasites are kept under control. The cleaner fish (*Labroides dimidarus*), for example, gets its food by gleaning the parasites from another fish, the wrasse (*Pseudodax moluccanus*). A wrasse (*Oxyjulis californica*) in California waters not only cleans various fish of parasitic copepods but even removes bacterial infections from the fish, which remains motionless during the grooming process. Some cleaner fish are even structurally modified so they can make "surgical" incisions to remove deep-seated parasites.

Some bottom fish in Florida associate with cleaning shrimps. The shrimps remove parasites and dead tissue from injured areas and even take bits of food from between the fishes' teeth with complete impunity. The California cleaning shrimp *Hippolysmata californica* also removes parasites and other organisms that attach themselves to the skin of fishes.

The benefits of commensalism are not always readily apparent. The horseshoe crabs, which are really not crabs at all, provide transportation for a varied passenger list, including barnacles, mollusks, and tubeworms. Perhaps the crab receives camouflage from this motley crew, which in turn is carried to new food supplies. The pea crab *Pinnotheres maculatus* is protected by living inside the shell of a scallop (*Aequipecten irradians*). Other crabs live inside oyster shells. There is no satisfactory answer to the question, "What does the scallop or the oyster gain from this association?"

The tribulations of the bean and Pismo clams of the California coast are even worse. A hydroid colony grows like a tuft of hairs at the end of the clam syphons. A tiny crab makes its home among the hairlike appendages of the hydroid colony, and still another crab lives within the shell of the bean clam. In the body of the clam itself, parasitic worms sometimes establish themselves. The razor clams of northern beaches also provide homes for some marine worms of the class Nemertea. And bay mussels have copepods in their mantles.

The innkeeper worm mentioned in Chapter 2 is probably the most frequently cited example of an animal that shares its home with uninvited guests. This sausage-shaped worm, a foot or more in length, lives in a U-shaped burrow with both ends of the burrow open at the muddy surface of western bay bottoms. Prey is caught in a net that filters microscopic particles from the current produced by movements of the worm's body. The worm periodically swallows the net and its contents, and then produces another net.

Conceivably, the worm could manage without its "guests," which include a scale worm (*Hesperonoe adventor*), a pea crab (*Scleroplax granulata*), a clam (*Cryptomya californica*), and one to a dozen or more arrow gobies (*Clevelandia ios*). Some cooperation, however, has been noted among the hotel inmates. If a particle of food lodges in the net and is too large for the worm to handle, the goby may take the morsel for itself. And if the meal is too large for the goby, it may take the food to the crab, which cuts it into smaller pieces and shares with the goby.

The forms of mutualism among sea animals are seemingly endless. We have seen that a hermit crab selects a comfortable snail shell to protect its unarmored abdomen. A worm also enters the shell and eats particles of food collected by the crab and possibly keeps the shell clean and free of other intruders. At the same time, the crab attaches a sea anemone to the top of the shell for camouflage and the protection afforded by the anemone's stinging cells. The anemone gets a ride to new feeding grounds as the crab moves with the tides. Some hermit crabs transfer their anemones to their new shells when they outgrow their old homes. Other hermit crabs use sponges or clumps of algae on their shells for camouflage.

In spite of all the symbiotic relationships, there are no VACANCY signs in the sea. Every available rock, sandy beach, frond of lichen, floating log, sunken ship, or piece of debris is support for some creature. Barnacles attach themselves to whales. A nemertean, *Polia involuta*, lives in the egg masses of crabs. Oysters attach themselves to pilings along Atlantic and Gulf shores. Surface films of bacteria and diatoms grow on eelgrass leaves in the bays. Hydroids grow on these films and then, in turn, are covered by velvetlike growths of algae, used by snails for grazing.

Even the casual visitor to the seashore soon becomes aware of the rhythms of the tides and seasons and of some of the creatures that are influenced by them. The grunion (*Leuresthes tenuis*), a little fish that spawns along southern California shores, is the classic example of an animal that is motivated by a biological clock. From April to August, grunion males and females come to sandy shores precisely at the turn of the high tide to deposit their eggs and sperm. They have only a moment in which to accomplish this task, for the next high wave covers them and they hurry back to the ocean. The eggs develop in the sand and, two weeks later, the next high tide washes the larvae back into the sea.

Along the Atlantic shore, the horseshoe crabs (*Limulus polyphemus*) crawl up from the ocean bottom and come ashore in late spring on the high tide, especially when the moon is full, to deposit their eggs and sperm in sandy nests. Two weeks later, the next high tide sets the young horseshoe crabs free and returns them to the sea. There they live and grow for several years until they are mature and able to repeat the process. These activities of the horseshoe crabs and the grunion are only two examples of rhythms of living things which are influenced by tides, the moon, the seasons, the light of day, and the darkness of night.

Periodicities affect plants, animals, and man in many ways. One of the mysteries of life is that we know little about these biological clocks. Many rhythms are linked with the rotation of the earth, which turns on its axis once every 24 hours and 51 minutes in relation to the moon. This is the period of many organismic fluctuations in the natural environment. These rhythms of approximately 24 hours, which include behaviors of sleepwalking, eating at specific times, and performing certain tasks, are entrained by the daily schedule of light and dark. They are called circadian rhythms.

Some rhythms tend to persist in the absence of the external stimulus, such as daylight and darkness, and require adjustment of the organism. Humans experience this when they encounter jet lag by flying thousands of miles in fast planes either toward the rising sun or away from it. While flying across the Atlantic Ocean to Europe, one experiences a very short night. I have left London at 12:00 noon and arrived in Los Angeles at 5:00 P.M. of the same day and have experienced the oddity of my hunger pattern drifting out of phase with my usual rhythm of daylight control.

This clock-controlled rhythm, which exists in practically all forms of life, including plants, is a subject of intense interest to biologists. Endocrinologists are also interested, since the role of hor-

Oyster shells are torn by storm waves from their attachments on stony bottoms. Once they are fastened to the substratum, oysters live their entire lives in one location unless they are moved by the force of the sea. Oysters frequently attach their shells to the shells of other oysters. This results in queer configurations as the combined shells grow larger.

mones in biological rhythms has introduced new dimensions into the study of animal and human behavior.

The clock, which may reside in the living cell, is influenced by outside signals from the environment and by internal physiological activities related to hormone production by the pituitary, pineal, adrenal, and other glands. It is well known that certain glands perform an important role in the periodicity of reproduction. Scientists are now finding that some of the secrets of biological rhythms that shoreline animals reveal may enable us to reset the clocks of destructive organisms in our human environment, thus bettering our lot.

Marine biologists and fishermen know that animal rhythms induced by the tides are important. Crab pots catch more crabs in bays during certain tidal movements. Oyster farmers provide collectors at the best tide level for the young oysters to attach themselves. Shore fishermen often follow the low tide far out on the beach so they can cast their lures into fish-inhabited waters. On the other hand, perch are best caught from onshore rocks when the tide is in, since these and other fish come near the shore when the tide is at flood stage.

Many other animals show rhythmic activity in tune with the tides. Gulls and many shorebirds arrive when the beach is exposed and small animals can be caught on the surface or probed from the sand or mud. Snails crawl to higher levels on rocks when the tide is high. Barnacles open their valves and feed. Clams push their siphons to the surface and pump water to obtain both food and oxygen. And fiddler crabs remain within their burrows. They also change color according to solar time, becoming dark in the day and light at night.

Occasionally one sees a cartoon showing a tiny fish about to be gulped by a larger fish, which in turn is being gulped by a still larger fish, and so on. Usually the cartoon is intended to depict something of a political or monetary significance. To a biologist, however, this sequence brings to mind the food chains, which are the basis for the maintenance and interdependence of all life.

The story of food chains, food webs, and energy exchange is absolutely characteristic of the seashore and the ocean itself. Marine food chains start with energy from the sun and the plankton of the sea. Trillions of diatoms and other simple plants convert the energy from light into protoplasm or living matter containing food energy, which feeds larger and larger organisms. The food chain may consist of only a few links: plankton to shrimplike krill to whale. Or it may include several links: plankton, the primary producers of stored energy; copepods, shrimp larvae, and other tiny organisms that graze upon the chlorophyll-bearing plankton; sardines; mackerel or other large fish; and, finally, sharks, seals, sea lions, and porpoises, the final carnivores.

Food chains are sometimes depicted as trophic levels that form a pyramid with a broad base and a narrow top. Organisms with identical methods of feeding are placed at various levels, with the green plants, or first trophic level, at the bottom. Only these organisms can convert the sun's energy into food by means of photosynthesis. The second trophic level included herbivorous organisms, those that feed on plants. A third level, the mixed-food group, includes many small animals, some of which feed on plants while others feed on very small animals. The fourth, fifth, and sixth levels include the primary, secondary, and tertiary carnivores. These are the levels depicted in the cartoon of the fishes swallowing one another.

Many food chains can coexist in a given community or ecosystem. When various food chains are interlocked they are called food webs. The web, or mesh, of food chains occurs because few animals rely on only one food source. The gull, for example, eats ghost crabs, which are a part of the shore food chain. It also eats fish, which are links in the marine food chains. The mole crab eats zooplankton in the sea but, in turn, is eaten by the sandpiper in the surf. Thus, many members of a community are bound together in a web of life and death.

Beach and dune webs are more readily observed by the beach walker than are the webs in the surf and beyond the sea. The high-tide mark on sandy beaches is frequented by animals with terrestrial affinities. But some marine species also inhabit or use the area at least part of the time. Here, land and sea food webs interlock.

Inhabitants or visitors to the high-tide drift line include tiger beetles, rove beetles, earwigs, and springtails. Scale insects and sucking bugs visit sea oats panicles when the grasses are in bloom. House mice and marsh rabbits also eat the seeds or the blades of grass. On southern Atlantic coasts, the ghost crab *Ocypode quadrata* visits the water's edge and lives in burrows on the high beach or in the dunes. It is both a scavenger and a predator. In season, the loggerhead turtles come out of the ocean, dig cavities in the sand of the upper beach, and lay their eggs. Then the sand crabs and the raccoons dig down to the nests and eat the turtle eggs or carry them away. Birds also must contend with predators from land and sea since the royal tern, least tern, American oyster catcher, Wilson's plover, and willet also nest in the sands of the upper beach.

The salt-marsh food web is characterized by two kinds of food chains, the grazer food chain and the detritus food chain. The primary producers of the grazer food chain are cordgrass and other herbaceous plants, which photosynthesize vast quantities of plant material, along with mud algae and phytoplankton, which also extract energy from the sun. These are grazed by various animals, including insects such as grasshoppers. The insects are preyed on by the clapper rail, the marsh wren, and the seaside sparrow, whose nests, in turn, are preyed on by rice rats, raccoons, and snakes.

In the detritus food chain, dead organic matter from plants and animals is decomposed by bacteria, fungi, insects, and even vultures. Crabs and snails also eat detritus in the salt marsh and are themselves eaten by the diamond-back terrapin. The fiddler crabs, likewise, are detritus feeders and are eaten by the red drum, willet, herring gull, snowy egret, and various herons. The importance of the detritus feeders and decomposers becomes apparent when we realize that all living organisms use the same materials over and over again. The metabolic processes of the decomposers break down dead plants and animals, excreta, and other waste products and liberate nutrient organic matter which once more becomes available to plants at the beginning of the food chain. Who knows? Maybe some of us contain a few molecules or atoms that once were in the body of a dinosaur.

When we think about food chains, there is much material for speculation. We, as humans, are products of radiant energy cycled upward through various trophic levels. Also, we are destined to be recycled just as the materials of life have been recycled for some billions of years. The food chain reminds us that on this earth there is no immortality in nature, on the land, at the shore, or in the sea.

The white sand dunes in New Mexico are composed of pure gypsum, not ordinary sand, like that found along the seashore. Plants grow mostly in the swales between the gypsum dunes.

6

The Dynamic Dunes

SAND DUNES ARE AMONG THE MOST ENCHANTING LAND FORMS along the seashore. These vagabond sand piles, between the sea and the forests of the coast, provide a lonesome landscape where you can explore mysterious swales and cranberry bogs and fish in ponds and streams that migrate with the whim of the winds. If you are an artist or photographer, there are fantastic graveyards of long-buried tree skeletons amid endless ridges of sand for your brush or lens. And, if you are energetic, you can climb a long slope, step over the cornice of a sand mountain, and ride the slip face of the dune down out of the screaming wind to sheltered valleys where the only noise you hear is the muffled sound of the distant booming sea.

There are many kinds of dunes in the world, and all are shaped by the wind. My first memories are of the "moving" or active dunes along the Lake Michigan shore and the "fixed" or stabilized dunes, covered with forests, inland from the shore. I walked their entire length from Gary, Indiana, to Three Oaks, Michigan, several times before the steel mills and the exploiters destroyed miles of their beauty. The American sea rocket (*Cakile edentula*) was there at the edge of the lake. I would see it later on Atlantic and Pacific shores. The beach grass (*Ammophila arenaria*) was also there, keeping the hummocks intact; around them, the sand was migrating to form foredunes, interdunal swales filled with reeds and sedges, and higher dunes where pitch pines, oaks, and eventually beach maple forests would conquer the sand.

There are dune wonderlands in our southwestern deserts, many of them of ancient origin. The colors of their sands vary according to the rocks from which they came: the Algodones dunes near Yuma, Arizona, are yellowish; pink and red dunes occur in the Utah deserts; the White Sands near Alamogordo, New Mexico, are made of dazzling pure-white gypsum blown by the wind from Lake Lucero, which in turn receives its gypsum from waters flowing from surrounding mountains.

These dunes, however, neither have all the moods nor offer the many diversions that intrigue visitors to the seashore. The history of most dunes begins in geological time, but the coastal dunes are unique because they are continually subject to the vagaries of the nearby sea. The coastal rivers still carry sand to the ocean. The shifting winds drift sand inland, and the rains and fog from the ocean cause plants to grow in unorthodox associations. Even the insect, bird, and mammal inhabitants of the dunes have strange ways that adapt them to life in a sandy ever-changing world.

Dunes are originally created by the action of waves and of wind, which moves the waves and ultimately transports sand grains inland. Above the shore, the dunes assume new shapes and motions in response to the restless shifting winds. Even the sand grains, originally abraded and rounded by the waves, are further molded, sorted, and redeposited as the dunes migrate, until they are captured by low vegetation and, finally, by forests remote from the open ocean.

Dunes of one form or another occur along most coastal areas with sandy beaches. In the northern portions of the Atlantic and Pacific shores, the dunes are of postglacial origin. The receding ice that once covered New England left hills of glacial till from Long Island to Cape Cod. As the ice melted, the ocean rose and the waves began the ceaseless process of grinding away the hills and ridges of till, scraping stones against stones and pebbles against pebbles until they were reduced to sand. Then the force of the winds began the process of heaping the sand into berms, or ridges, near the shores. From these the grains were picked up and built into hills and valleys that move as the prevailing winds shift with the seasons.

Some of the dunes, such as those at Provincetown and at Sandy Neck in Barnstable on Cape Cod, are rejuvenated dunes set free by man and his timber-cutting activities in recent times. The grazing cattle of early settlers destroyed the sand-binding grasses of other dunes. Fires that burned the trees also laid the landscape bare and set the hills to marching again.

The Outer Banks of North Carolina and the coastal islands of Georgia and the Gulf coast also had their beginnings in geological time. As the glaciers melted, the sea returned to a higher level. Waves and storms then built the sand, formerly deposited by rivers, into reefs and islands. As the wind-blown sand moved toward the mainland, these islands lost their sand on the ocean side and were built up on the landward side. Many of these islands were stabilized by grasses near the shore. Farther inland, shrubs, pine, and live oak occupied the uplands where rainfall had leached the salt from the soil. But, as usual, man has started the sand moving again by cutting trees, grazing the grasses with livestock, building beach houses, and constructing roads and boat landings.

The Oregon seacoast dunes, between Coos Bay on the south and Florence on the north, are among the most fantastic sandy lands in America. For dune explorers and fishermen, these are a veritable paradise of smoking ridgetops, flower-strewn swales, wandering lakes and ponds, and erratic streams. There are big lakes 4 or 5 miles from the sea and small lakes and ponds in the moving dune strip that borders the ocean.

The variety of fish in these lakes seems endless. Bluegills and largemouth bass are plentiful, as are crappies and catfish. When you believe you have hooked a bass, it may turn out to be a native cutthroat trout. Yellow perch and steelhead are where you find them. Silver and Chinook salmon come up from the sea to spawn. While fishing for silver salmon in Tenmile Creek, I once caught a flounder. Where else in the dunes of America can you catch a flounder?

The dynamics of the dunes begins at the beach front. Mild summer waves build berms near the tide mark, and the winds move the sand inland to form dunes. The beach itself is inhospitable to most plants and to many animals. Those that burrow, such as mole crabs and razor clams, find a suitable environment near the water's edge. Microscopic animals and a few small creatures that practically "swim" in sand also live in the harsh environment of the sea front. Sea turtles and birds are temporary visitors that nest on the beach. These animals do not contribute to dune formation. Only plants adapted to life in the shifting sands perform a vital function in the beginning of dune formation.

The first dune, or foredune, back of the shore generally owes its origin to sand-binding plants with runners and rapidly growing root systems. We have seen that sea oats (*Uniola paniculata*) and beach grass (*Ammophila arenaria*) are typical dune stabilizers. These grasses disrupt the flow of air and allow sand grains to settle on the leeward side of the plant. Soon, the mound of sand itself traps more sand and the foredune grows, sometimes to a height of several feet in a single year.

Sand that blows over the foredune accumulates in other dunes to form a series of parallel ridges. The largest dunes are farthest inland. Some reach such heights that the wind no longer transports sand over their tops, and these dunes become stabilized by grasses, herbs, shrubs, and forests. The windward slope of dunes usually is gradual while the leeward side is steep. Sand blown over the crest of the dune slips down this face and invades the valley between it and the next dune. The wind also moves sand from this valley until it reaches moist sand near the water table. If this area is extensive, it is called a deflation surface or a deflation plain. It is a favorable place for many herbaceous plants to grow.

Many coastal dune systems show a mixture of patterns other than parallel ridges. In summer, on the Oregon coast, small dunes develop as transverse ridges at right angles to the northwest winds. These dunes are partly destroyed by the winter winds. Unique to the Oregon coast are large dunes, 150 feet or more in height and up to a mile in length, which develop slip faces on alternate sides with seasonal changes in wind direction. These "oblique ridges," named by William S. Cooper, are oriented obliquely to both northwest and southwest winds and are primary invaders of dune forests.

Blowouts develop when the vegetation cover of coastal dunes is disturbed by grazing animals, automobiles, or even by people on foot wearing paths in the sand. Storm waves from the sea also may cut away part of a foredune, leaving a valley and cliffs of loose sand. Such a blowout can migrate over other dunes as it develops into a parabolic U-dune with an advancing front of loose sand and trailing arms, which may become partially vegetated by sand-binding grasses. Parabolic dunes occur on both the Atlantic and Pacific coasts.

Many miles of the Atlantic and Gulf coasts are fronted by narrow elongate islands which owe their origin and maintenance to sand dunes. These barrier islands grow and move with the shifting sands. The dunes that form them stop storm waves from cascading across the landscape and destroying stabilized dunes, forests, and marshes behind the last row of permanent dunes.

The large-headed sedge with its dense terminal head inhabits areas of sand movement along the Pacific coast. It grows in company with seashore bluegrass and various herbs with colorful flowers in moist meadows and dune swales.

Barrier islands typically support a variety of natural systems, including grass communities, marshes, slough vegetation, understory shrubs, deciduous forests, coniferous forests, and mangroves. These offer life support for birds, mammals, fish, insects, and other numerous invertebrates. They also contribute nutrients to the coastal ecosystem.

If one traverses a barrier island from the sea to the inland side, numerous stages of ecological development may be observed. Beginning at the ocean beach, the berm is succeeded by the shifting dunes and these, in turn, by the stabilized dunes, held in place by shrubs and grasses. On south Atlantic beaches, the sawtooth palmetto also contributes to stabilization of the sand. On the island uplands are forests, where seabirds breed in rookeries, and sloughs, where fish and other wildlife abound. On the leeward side of the island are marshes and tidal creeks, where the fish spawn and clams, shrimps, and other marine species spend the early stages of their lives.

Nowhere can one see nature in action better than on the traveling dunes. Under the assult of the sea winds, sand is whisked from every surface and inexorably moved inland to battle with the grasses, herbs, shrubs, and trees. The first stages of stabilization begin with clumps of beach grass that slow the progress of the wind and sand. In the swales between the dunes, beach heather, bayberry, and beach plums carpet the surface and shield the sand which otherwise would be removed by the wind. Finally, the sand is covered by wooded hills far to the rear. But even these are not always exempt from invasion when rogue dunes break loose and pile their sands along an advancing front that covers trees and all other things in its path. Eventually, the sand may move again and the trees will be resurrected as skeletons in a stark and fantastic landscape.

The plants of the dunes become even more interesting when we consider the characteristics that enable them to grow in the harsh environment of the beaches and the dunes. Beach plants must survive salt spray, constant wind, high temperatures, and intense sunlight. Plants of the foredunes must grow rapidly and have fibrous root systems in order to cope with sand accumulation and to bind sand into place.

In protected areas behind the dunes, especially in marshy areas, herbs must have the capacity to germinate, mature, and produce seeds before the sands dry out in summer. Shrubs and trees on the high dunes must endure sandblasting and pruning by the wind. They do this by vigorous sprouting. If the windward sides of shrubs and trees are killed by sandblasting and salt spray, the plants must be able to survive by growth on the leeward sides of trunks and branches. This type of growth accounts for the asymmetrical forms of many woody species.

Even in one hour of wandering over the dunes you can see many examples of plants that have adapted to survive. Annuals such as the saltwort or Russian thistle (*Salsola kali*) and camphorweed (*Heterotheca subaxillaris*) grow quickly, colonizing dunes in one place in a given year and in another place the next year. The railroad vine (*Ipomoea stolonifera*) and bearberry vines (*Arctostaphylos uva-ursi*) form mats of stems and leaves that hold down the sand. Plants with leathery leaves retain water, as do the prickly pear cactus and the succulent sea fig (*Mesembryanthemum chilense*), which grows so abundantly on California's coastal sandy areas.

Only the hardy pioneers, in particular the beach grass (*Ammophila arenaria*) and the sea rocket (*Cakile edentula*) are prominent on foredunes. In the central dunes, the number of species increases. And in the hind-dune area, the diversity of species is pronounced. The overall result is a zonation of vegetation from mean high tide to interior forests or shrub lands. This zonation runs more or less parallel to the coast.

The sea rocket has always intrigued me because of the demanding habitat in which it grows, right on the beach sand. I first saw it on the southern shore of Lake Michigan while on a field trip with Dr. Henry C. Cowles of the University of Chicago. He called it a true pioneer but offered no explanation as to why it succeeded on the beach and not in the apparently more favorable sites in the interdunal swales.

Part of the explanation came years later from an experiment at Bodega Head, California. Scientists planted seeds of the sea rocket (*Cakile maritima*) on the strand, in unweeded grasslands, in weeded grasslands with strand sand, and in weeded grassland with grassland soil. Seeds in all four treatments germinated, but growth in the grassland plots was poor. Laboratory tests indicated that sea-rocket seedlings need a high light intensity, 3,000 or more footcandles, which they do not receive in the shade of grasses and other plants.

The sea rocket grows on both Atlantic and Pacific shores. Its four sepals and petals distinguish it as a member of the mustard family. It is an annual, fleshy, branched plant with leaves pinnately cleft. The variety *C. edentula* var. *californica* is native from San Diego to British Columbia. The fleshy two-jointed fruits are carried by the waves to the upper beach, where their seeds sprout and grow each season.

The beach grasses are among the most important of the sand-binding grasses, for they help create the foredunes. There are two kinds of beach grass: European beach grass (*Ammophila arenaria*), introduced from Europe and now common on dunes from California to Washington as well as Cape Cod, and American beach grass (*A. breviligulata*), native to sand dunes from Newfoundland to North Carolina and on the shores of the Great Lakes. Other names applied to these grasses are marram, psamma, and sea sand reed. The genus name, *Ammophila*, comes from the Greek *ammos*, sand, and *philos*, loving, alluding to the habitat. The specific name *arenaria* also means sand. Freely translated the whole name means, "I love the sandy sand."

The beach grasses are strongly adapted to life in a harsh environment. Their stiff stems rise from rugged rapidly growing rhizomes, which penetrate the sand and give rise to fibrous roots that anchor the plant and provide moisture from subterranean layers in the dune. When sand covers the leaves, the underground stems elongate, rise to the surface, and produce new leaves. The leaves also adjust to temperature and humidity. On dry sunny days, the leaves roll inward to reduce evaporation. On rainy days or when the fog is in from the sea, they unroll and are almost flat. The leaf blades are stiff and sharp-tipped and well able to withstand the wind. The rigid plumes of flowers and seed heads add to the picturesque scene of the oceanside as they sway before the ever-shifting winds.

Another sand-binding plant is beach heather (*Hudsonia tomentosa*). It also has the inappropriate name of poverty grass. Beach heather is neither a grass nor a true heather; rather, it is a member of the rockrose family. It grows in mats close to the sand and forms patches of sage green with its oval or oblong leaves, which are hoary with down. In autumn, these patches are tinged with yellow, and in

The great sand dune east of Provincetown, Cape Cod, still moves as winds come in from the Atlantic Ocean. Sand grasses have stabilized the dunes only in local areas.

winter they fade to a sandy gray. In June, beach heather paints the dune sides yellow with its showy golden blossoms. Its relative, golden heather (*Hudsonia ericoides*), is also downy but more greenish, and its flowers grow on slender, naked stalks.

The beach cocklebur (*Xanthium echinatum*) grows near the edge of the sea. Its burrs are thicker and plumper than those of its relatives, which I was acquainted with in the cornfields of the Midwest. They filled our horse's tails until they resembled prickly clubs. The body of the burr is mostly pubescent, and even the hooked prickles are softly hairy at their bases.

The seaside spurge (*Euphorbia polyganifolia*) spreads itself in lowly mats throughout the dunes. Its prostrate growth protects it from the winds, and its central taproot draws moisture from deep in the sand. Its leathery leaves also reduce transpiration. Like the snow-on-the-mountain (*E. marginata*), which grows in my Colorado garden, the seaside spurge and its relatives have roots and stems containing sticky milky juice. This alone distinguishes them from some of the cacti that otherwise resemble them.

At the top of the foredune and on the border of salt marshes, the seaside goldenrod (*Solidago sempervirens*) is at home with the beach pea (*Lathyrus maritimus*). This goldenrod differs from the more than fifty species of its kind because of its ability to endure salt and its succulent, water-storing stem. The dead dried plants stand foursquare against all the winds that blow across the dunes in winter.

Beach peas are common on the foredunes; you can see their stems meandering over the sand, and their plentiful purple flowers and small peas all summer. The six to ten leaflets, the large purple flowers, and the stipules, nearly as large as the leaflets, distinguish the beach pea from its other relatives in the genus *Lathyrus*.

Various shrubs find shelter from the winds in the sandy realm back of the foredune. Here are bayberries (*Myrica pennsylvanica*), beach plums (*Prunus maritima*), wild cherries (*P. serotina*), and bearberries (*Arctostaphylos uva-ursi*). Bearberry vines cover the ground on level sandy areas and make carpets beneath the pitch pines behind the secondary dunes. These adaptable plants with their bright red berries are widely distributed. They grow on rocky hills in New Jersey and Pennsylvania and northward and also flourish on dunes and in the forests in the Rocky Mountains and in the Pacific Northwest.

On stabilized dunes, pitch pines (*Pinus rigida*) and scrub oaks

(*Quercus ilicifolia*) grow in stunted forms, somewhat like the Krumholtz or distorted spruces in the subalpine zone of the Rocky Mountains. In the mountains, wind and snow shear off the tops; the trees persist only because they grow behind sheltering boulders. In the dune country, it is now believed that salt spray deforms the trees by killing the growing tips of branches.

If you explore Atlantic dunes you will find other trees, including white birches, clumps of alders, and stands of aspens and willows. You will also find moist swales and bogs that house a wealth of attractive herbs and shrubs. The cranberry vines may have intermingled with blue irises (*Iris versicolor*). And in early summer two beautiful orchids may be found in bogs: the rose pogonia (*Pogonia ophioglossoides*), a bearded orchid with a usually single magenta-pink or, rarely, white flower, and the grass pink (*Calopogon pulchellus*), with a raceme of half a dozen or more magenta-crimson flowers and a solitary, grasslike leaf.

A rich variety of dune vegetation is found along the edge of the Pacific Ocean. The usual communities of sand colonizers and stabilizers are present but the plant species are mostly different from those of the Atlantic and Gulf shores. The dunes of California, Oregon, and Washington also differ from eastern dunes because of the belts of almost impenetrable shrubbery that border them. In the northern reaches of the Pacific shore, lodgepole pine (*Pinus contorta*) and Sitka spruce (*Picea sitchensis*) come down to the shore, where they are desiccated by salt spray and abrading sand. Where the dunes become stabilized, gigantic Sitka spruce and Douglas fir (*Pseudotsuga menziesii*) dominate.

Among the plants you are sure to see along the Pacific shore are the sea rocket and the coast strawberry (*Fragaria chiloensis*). The former grows nearest the reach of the tide, the latter on sandy areas above the tide mark, on stable dunes, and in the moist interdunal swales. The strawberry produces long stolons like those of garden strawberries. Another easily recognized plant is the coast morning glory (*Convolvulus soldanella*). Unlike other members of this genus, it does not grow in long creeping or climbing vines, but spreads out to form a prostrate mat of glossy green foliage.

The dunes support a host of other early colonizers, but do not expect to see them all in one place. Part of the pleasure in finding them is to note the slight differences in habitat to which different species are adapted. Seashore bluegrass (*Poa macrantha*), for example, grows on beaches and dunes, thriving in the moister sites. On the foredunes, European beach grass grows in association with dune wild rye (*Elymus mollis*).

Dune visitors, especially in California, are sure to see three members of the carpet-weed family. The Hottentot fig (*Mesembryanthemum edule*), an introduced species, is distinguished by its three-sided succulent up-curved leaves. Its yellowish flowers reach a diameter of 3 or 4 inches. The sea fig (*M. chilense*) has similar leaves but they are not curved; its rose-colored flowers are 1 or 2 inches broad. The third succulent species, the ice plant (*M. crystalinum*), has broad leaves covered with crystal-like vesicles that give it an icy appearance. The ice plant, an annual, occurs from Monterey Bay southward to Baja California. All these plants produce showy flowers and, because they are excellent soil binders, are often planted along roads to help control erosion.

Some of the members of the heath family come out to the coast on sandy areas and some have strikingly beautiful flowers. The western azalea (*Rhododendron occidentale*) is a deciduous shrub with leathery leaves and white flowers that sometimes show a pinkish cast. It grows from southern Oregon to northern California.

The salal (*Gaultheria shallon*) is a straggling shrub that sometimes forms impenetrable tangles on sandy oceanside terraces. Its evergreen leaves are oval or oblong, narrowed at the apex and dark glossy green above. Its white or pink urn-shaped flowers are borne in loosely flowered racemes, 3 to 6 inches long, and its fruits are dark purple or black with protruding styles. Salal is especially abundant immediately north of Lincoln City, Oregon.

The greatest variety of dune plants occurs on the moist sands of the deflation plains on the inland side of the foredunes. These plains are depressions between the dunes where sand has been blown from the windward slopes until the surface is near the water table. This produces habitats where various meadow and marsh communities can flourish. In these areas, when the sand becomes dry in summer, the seashore lupine (*Lupinus littoralis*) is common. Associated with it is beach silver-top (*Glehnia leiocarpa*), a member of the carrot family. The leaf bases of this plant are almost buried in the sand, giving it a rosette appearance. The leaves are pinnate and the leaflets are densely woolly on the lower surfaces. The inflorescence is compound, consisting of numerous umbels, which grow as a compact globose clump and yield corky-winged fruits, each with ten wings.

Moist meadow communities in the dunes usually support a dense growth of many species. Some are evanescent and inconspicuous. One of these is little hairgrass (*Aira praecox*), an annual that dries up in early summer. Other grass and grasslike species are red fescue (*Festuca rubra*), brown-headed rush (*Juncus phaeocephalus*), and sickle-leaved rush (*J. falcatus*). Spring-bank clover (*Trifolium wormskjoldii*) grows here and in seepage areas on banks and cliffs above the sea. The flowers vary from white to red to purple and are about ½ inch long.

In very wet meadows and marshlike communities, one can find such interesting plants as golden-eyed grass (*Sisyrinchium californicum*), twisted orchid (*Spiranthes romanzoffiana*), and sundew (*Drosera rotundifolia*), a rosette plant with leaf blades covered with sensitive glandular hairs which trap insects.

The strange "cobra orchid" (*Darlingtonia californica*) grows in some of the dune bogs. It is a carnivorous pitcher plant, found nowhere else in the world except in the dune country between Coos Bay and Florence, Oregon, and in a few places in northern California. One other interesting plant of the marsh community is bog club-moss (*Lycopodium inundatum*). This plant reproduces by spores instead of seeds.

The seashore visitor can find many diversions in the dunes. There are strange plants to discover, the tracks of dune animals to study, and ripple marks and wind-blown scenery to photograph. Camp meals can be cooked over driftwood gathered on the beach. In some areas, particularly along the Oregon coast, you can take exciting rides in dune buggies that roar up the inclines and drop precariously over precipices of sand. These balloon-tired machines are not harmful if their use is restricted to areas of open, unvegetated sand and if they do not destroy sand-binding plants that control dune movement.

The dynamics of the dunes, ranging from the small to the gigantic, are always cause for fascination. If you lie in the shelter of a dune hummock you can watch sand grains leaping past, some traveling so fast they cannot be seen by the human eye. This motion is called saltation. In a high wind, the grains leap and bound an inch or two off the surface and strike the other grains like minute billiard balls. The impacted grains advance so that below the filmy sheet of bounding grains the surface sand also creeps. If you stand below the cornice of a large moving dune you can see the peaks smoke like chimneys. The sand grains fall on the leeward slope and roll down the smooth slip face of the dune.

Dune topography moves in a small way beneath one's very feet. Ripple marks are like lilliputian dunes on big dunes. They form parallel ridges when the wind blows. Like the large dunes, they slope gently to windward and steeply to leeward. In a strong wind, they migrate and form ever-changing patterns. Years ago at Ipswich, Massachusetts, on a blustery March day, Charles Wendell Townsend watched some ripple marks that were 4 inches apart from crest to crest. He observed them advancing at the rate of 1 foot in eight and a half minutes.

The dunes can be gentle. When the Gulf islands are not racked by squalls, you can explore their dunes, see eagles nesting in the

The pitcher plant, or "cobra orchid," is an insectivorous plant found in sphagnum bogs along the inner dune country of the Oregon coast. It grows only from Lane County, Oregon, to northern California—nowhere else in the world.

trees, hear alligators bellowing in the swamps, and follow rabbit tracks over the sand. But when the hurricane roars, even the stablized dunes surrender to the wind. In 1969, powerful hurricane Camille cut Ship Island in two and ripped away a third of its vegetation.

The full exhilaration from a visit to the dunes comes in winter, when the storms fiercely drive the sands, the gulls scream overhead, and the water birds stream endlessly over the restless foaming sea. Then the scent of the marsh reaches your nostrils and the loneliness of the scene fills you with a sense of fulfilling solitude. But eventually, the storms pass and the dune environment again becomes a place for leisurely excursions and the rediscovery of how living things exist.

There are animals on the dunes. On calm sunny days, the only moving living creature may be the seaside grasshopper (*Trimerotropis maritima*). As you walk across the sand it will apprise you of its presence by jumping, flying, and making a crackling sound just before landing. To observe it closely you will have to watch where it lands, because its pale color blends with the sand. Some are nearly pure white. Others are mottled or banded with white for almost perfect camouflage.

A dune dweller that needs no camouflage is the velvet "ant" (*Dasymutilla bioculata*). This creature is not an ant at all, but a parasitic wasp. Do not pick it up; its sting is excruciating. The furry red coat is highly visible and undoubtedly is a warning to creatures that might mistake it for a morsel of food. It lays eggs in digger-wasp nests where its young can feed on the digger larvae.

The ant lion larva is easy to locate by its cone-shaped pit in the sand. This fierce creature lies in wait for unwary ants and even spiders to fall into the sicklelike jaws that project from the bottom of the sand funnel. To speed its prey down the side of the pit, the ant lion larva tosses sand with its head, causing the sides of its pit to cascade to the bottom along with the luckless victim.

Ant lion larvae may be kept in a jelly glass partly filled with sand or sugar. They will excavate a pit in short order and will eat flies, ants, other insects, and spiders you find in or around the house. They will maintain their pit and live a long time without food; I have kept them for several months. But the reported record for an ant lion larva to go without food is 240 days. The ant lion, also called doodlebug, pupates in the sand and eventually emerges as a lace-winged adult that mates, lays eggs, and then dies.

Tiger beetles are among the most active insects on the dunes. They run over the sand incessantly in search of insect prey. Their eyesight must be exceptional. Whenever I attempt to photograph them in their natural habitat, they turn and look at me and then run just when they seem to be in range of the lens. Their beautiful metallic green, bronze, and other colors make them excellent subjects for color photographs.

Spiders also are common on the dunes. Some of the wolf spiders have a leg spread of an inch or more. These spiders do not spin orb webs but hunt, using their eight jewellike eyes to locate insect prey. They are generally pale in color and blend with the sand. In the daytime, you can find their nesting places by looking for pencil-sized holes. Wolf spiders carry their young on their backs while hunting.

As in the ocean, animal life on the dunes exhibits a variety of food chains. Many insects subsist on green plants. They, in turn, are captured by insect predators, birds, and toads. The toads are eaten by hognose snakes and the snakes are captured by hawks.

White-footed mice, deer mice, jumping mice, shrews, and cottontails indicate their presence by the tracks they leave on the dune sands. The tracks of foxes, skunks, and raccoons indicate that these larger mammals also make marauding visits to the dunes, as do the marsh hawks and owls that come from the shrub thickets and nearby forests. All are a part of the food web of the dunes.

Relatively few mammals inhabit the Pacific coast dunes. Deer mice are restricted to burrowing in stablized sand. Foxes, raccoons, and weasels occasionally appear on the sands, but they make their homes in adjacent wooded areas. Deer, jackrabbits, and an occasional bear wander over the dunes.

Kangaroo rats inhabit the dunes, emerging from their burrows only at night. These appealing creatures are exceptionally adapted to dry environments. They never have to drink since they make metabolic water from the seeds they eat. Their urine and feces are highly concentrated, so they use up only a little water when they excrete. In addition, their nocturnal rhythm conserves moisture that otherwise would be lost during the heat of the day. Even their burrows are closed in daytime to conserve moisture.

Birds are numerous in the dunes during certain seasons. Gulls sometimes use the dunes for nesting sites. From New England to Texas, fish crows scour the sands for carrion in winter and for bird eggs, insects, and wild fruit in summer. Where sea oats abound, song sparrows and redwing blackbirds forage for seeds. Savanna sparrows nest at the edges of grass clumps, and flickers sometimes nest in dead trees that have been uncovered by the shifting sands. On Georgia and Carolina dunes, other nesting birds include Wilson's plover, willet, royal tern, least tern, and the American oyster catcher. The black skimmer also nests in sandy areas.

However, some birds are rarely seen in the dunes. Count it a banner day in winter if you see a snowy owl sitting on a dune top watching for rabbits. This great white owl migrates southward along Atlantic and Pacific shores only when the lemming supply in the Arctic, its natural habitat, is low.

In fall and winter other birds are present in the dune thickets and in the pine groves. The red crossbills and the white-winged crossbills hang from pinecones and extract their seeds. In the dune woodlands, you may also see the pine grosbeak and the common redpoll, which spends much of its time on the ground searching for seeds in weedy areas. Autumn in the dunes is the time to see warblers on migration. In the natural setting of sea, sand, shrublands, and woods, dunes are the birdwatcher's delight.

Toads are common inhabitants of dunes and upper sandy shores. Their tracks in the sand indicate where they have searched for insects.

7

Estuaries: Where Two Water Worlds Meet

RECENTLY, I TOOK A SAILBOAT CRUISE on that largest and most complex estuary of the United States, Chesapeake Bay, where sky and water dominate the scenery. It was a tranquil day. The breeze was gentle, the waves were low, and the auxiliary motor of our trimaran came to life each time we needed additional speed to avoid the mighty cargo ships steaming up the bay. We boarded at Solomon's Island, sailed past Thomas Point Light, northeast to the eastern shore, and then southwest by West River, where the crab pots are. Next, we passed between West River and Herring Bay by Bloody Point Light and back to Harness Creek off South River. In that whole day we did not see a thousandth part of all that can be seen and experienced on this queen of estuaries. But the serenity of the cruise gave me time to consider some of the things I knew about the prodigality of plant and animal life in estuaries where river water meets that of the sea.

The myriad of living things in estuaries is made possible by the wide range of habitats produced by tides, variations in salinity, presence or absence of sand and mudbanks, and the relation of light to photosynthesis by algae and other plants.

Algae, as we have mentioned, are the organisms at the bottom of food chains. They lack higher neural functions, as most plants do, and, as plankton, they are forever bound to involuntary simplicity. From their viewpoint, if any, they have no purpose in life. They have no romance since they reproduce by simple cell fission. They

do not deliberately visit among neighbors and forests in the fore-
ground. They do fluctuate in number, as living conditions for their
existence are changed by forces beyond their control. But their persistence in quantities almost beyond human comprehension fosters
the chain of life that extends upward with increasing evolutionary
complexity. Consider the creatures that depend on the micro-
plankton.

   The food chains and webs in estuaries included clams, scallops,
oysters, snook, mullet, menhaden, sea bass, and other important
commercial species of fish. Teeming birdlife ranges from sandpipers, curlews, gulls, ducks, geese, and whistling swans to owls and
eagles. These animals do exhibit a complexity of purpose and action.
When thousands of dunlins pour into an estuary in winter, the
whole flock seems to move with a common intent. Their primary
goal, of course, is to feed on organisms that started with plant life in
the waters of rivers, the estuary, and the sea. The actions of these
different and more highly advanced species vary according to their
navigational abilities and their individual food requirements. Fish,
for instance, move independently or in schools. Ducks, geese, cormorants, and pelicans fly in formations, each according to its own

Young great blue herons brought down from their nest for photographs. As
many as five sizes of baby herons may live in the same nest, since the eggs do
not hatch at the same time.

species. Sanderlings behave like movable living mats motivated by waves on the beach. I recently asked John K. Terres, who has written so much about birds, which bird decides to change the course of the flock. He said he did not know. This is only one of the mysteries one can readily see on estuaries.

Estuaries are many things to many people. The biologist thinks of an estuary as the outlet of a river whose water has a variable salinity owing to its contact with the sea. This gradient of dilution of seawater produces a range of habitats for plants and animals that require environments varying from terrestrial to marine. Technically, an estuary is any confined coastal water body with an open connection to the sea and a measurable quantity of salt in its waters. In a seven-volume report (*National Estuary Study*) submitted to Congress on January 30, 1970, an estuary is further described as follows:

An estuary is wings whispering over marsh-fringed waters; a camouflaged nimrod anticipating duck dinner; muskrats scurrying through reed, sedge and grass; shorebirds feeding busily on a tidal flat; brant noisily foraging in eelgrass; oystermen silently tonging the bars on a misty morning; a rod fisherman jubilating over a prized striper; sea oats bending in the ocean's breath on a lonely dune; headlands against fleecy clouds and ocean surf; sailboats wending through gentle swells, stark in the sunset. Yes, it is all these and many more. It is commerce, it is harborage, port facilities, and fringing industry and homes. It is a gigantic energy converter. It is succulent oysters, toothsome shrimp, tender flounder, fodder supplements for poultry and livestock, organic medicinals, agar, oil, and sulphur, even gravel and still more. And it is beauty and solace; yet a scientist's laboratory.

At their best, estuaries are clean attractive places where enormously important aquatic plants grow in shallow waters. They support marshes that filter sediments washed in from the land. They are reservoirs for nutrients that give life to large numbers of invertebrates, fish, reptiles, birds, and mammals. Their shellfish, shrimp, and crab fauna are harvestable products. If one observes closely, there is more vibrant and varied life in an estuary than in any other part of the seashore.

Nearly 900 estuaries exist along the shores of the United States. Among the great ones along the Altantic Coast are the Chesapeake and Delaware bays, both drowned river valleys that were flooded by the last glacial melting. Farther south, along the coast to Florida and the Gulf of Mexico, are many coastal lagoons that are partially enclosed by barrier beaches and sandy islands. These lagoons are enriched by rivers that mingle their waters with those from the open sea.

The most fabulous estuary on the Pacific coast is San Francisco Bay. It is the end result of earthquakes, mountain building, and sea-level changes. Seawater enters this estuary through the Golden Gate and fresh water comes from the Sacramento–San Joaquin delta and the Suisun and San Pablo bays, miles inward. San Francisco Bay is not as large as Chesapeake Bay, but it has an unusually wide variety of environments and offers many recreational possibilities.

The area of this estuary that is subject to tidal action extends to the delta, a tidal marsh watered by the Sacramento and San Joaquin rivers. This area consists of some forty islands, which are now farmed, surrounded by nine-hundred miles of navigable waterways. This once magnificent natural environment is now intensively polluted and teeming with docks, commercial vessels, sport boats, waterfowl hunters, and commercial fishermen.

Chinook salmon (*Oncorhynchus tshawytscha*) are native to this estuary. They spawn in the tributary rivers in such numbers that San Francisco Bay has one of the largest chinook salmon runs in the world. The introduced striped bass (*Roccus saxatilis*) also is plentiful here; its annual sport catch has been estimated at between 1.4 million to more than 2.5 million fish, most of which stay within the estuary. Some of the wildest fishing I have ever had was for stripers, at the very edge of the south pier of the Golden Gate Bridge. Coupled with the thrill of a strike and battling the fish up from the depths was the fear that the captain could not avoid capsizing the boat against the concrete pier. The outgoing tide and waves lifted and lowered the boat many feet while we were within arm's reach of the mighty column that extends skyward to the bridge.

How different this dynamic estuarine opening to the sea is in comparison with many of our smaller estuaries. The Salmon Creek estuary in the Sonoma coast region of California is still bordered by marshes and a variety of freshwater and saltwater habitats. There is a small salmon run in the Finley Creek tributary. Tomales Bay includes a large area of intertidal marshes and mud flats as well as an interesting population of Japanese invertebrates and other exotic species, introduced through years of oyster culture. The Elkhorn Slough system, between Pajaro and Salinas valleys, is one of the larger estuaries in California. It too is threatened by industrial and harbor development.

Some estuaries, where the river flow is large and seasonally variable, produce no marshes or extensive mud flats. The Columbia River, for example, has moderately strong tides that reverse the current for many miles upstream. But large volumes of sediment are carried into the Pacific Ocean, resulting in bar formation. Anyone who has crossed the bar in a small boat amid high waves knows the relief that comes upon finally emerging into the still mighty but gentle swells of the broad ocean itself.

River inlets to estuaries provide some of the best sport fishing in coastal waters. On one fishing expedition I drove with my two sons out of Coos Bay, Oregon, over the Isthmus Slough, skirted the Coos River, and ended on the North Fork where the stream narrows down to a boat canal. The water was mirror smooth when we arrived. Just as we finished assembling our fishing gear and tying on yellow pippins with thin hackle and red paint at the ends, the water began to move upstream. The tide was coming into the bay.

With several other fishermen, we got our lures into the water. Immediately, we had shad on the hook. The fish flipped into the air, gyrated at the ends of our lines, and sliced back into the water like the fiercest trout I have ever battled in mountain waters. The fun lasted for an hour. Then the tide ebbed and the fishing was done.

That same day we rented a boat and fished for flounder in Siletz Bay at ebb tide, using ghost shrimp for bait. A flounder at the end of the line feels different from any other bay fish. It begins with a few nibbles, followed by a strong tug. When you set the hook, the fish flaps along beneath the surface like a tailless kite in a stiff breeze. The pleasure of pulling in these flapping porcelain-white dinner-plate-sized fish is ample reward for a few hours spent on the rich coastal fishing grounds of a calm estuary.

Many people are unaware that estuaries and other wetlands are among the most productive natural ecosystems on earth. An estuarine salt marsh can produce many times the amount of biological material that can be grown in an Iowa cornfield or Appalachian forest grove.

Estuaries are welcome mats for animals that live their beginning stages in mixtures of fresh and tidal waters. At some point in their life cycles, some 60 percent of all commercial marine fishes and invertebrates rely on coastal wetlands. Two thirds of the fish species of the Atlantic and Gulf coasts are dependent on these waters, as are oysters, clams, and mussels. Shrimp and crabs use them for nurturing their young. Many birds find food and nesting sites in the edges of these river outflow areas, and migratory birds find winter refuge in estuarine potholes and marshes.

The biological significance of estuaries is of concern not only to animal and plant life but also to people. Estuaries are protective natural systems that can counteract human exploitation of the land. Estuarine marshes and waters act as enormous filtration systems that absorb impurities and pollutants from the water. They equalize water flow and protect the coasts from freshwater flooding, ocean storms, and hurricanes.

If one attempts to list all great and small plants and animals in estuarine waters, the count seems endless. The fish life alone is astounding; species in a single estuary may number in the hundreds. Some, such as fluke, bluefish, menhaden, and king whiting, spawn in the open sea, and their tiny young drift shoreward and seek refuge and food in shallow estuarine waters. In contrast, migratory fish such as weakfish, redfish, mullet, and black drum, spawn in estuaries; their young grow up in shallow water and eventually migrate to the sea. And the spotted sea trout of South Atlantic waters resides in estuaries year-round.

The biology and the subtle mystery of estuarine marshes and mud flats lack the dramatic appeal of waves dashing against rocky cliffs or the enticement of sand on a level beach. But not to a confirmed clam digger. The bivalves—clams, oysters, and scallops—have many modes of life and provide food and enjoyable sport for one who knows how to obtain them.

Some bivalves burrow in sand, mud, rocks, or wood; others are free-living and are able to swim short distances. The purple clam lives in estuary entrances. Gaper clams and littlenecks prefer the quieter waters of bays. Bay mussels attach themselves to pilings and roots of marsh grasses. The rock scallops, jingles, oysters, mussels, and chamas are exposed and easy to find. Bean, Pismo and the common littleneck clams bury themselves a few inches below the surface. The thin-shelled littleneck, purple clam, and most of the piddocks go to a depth of a foot or more, while the gaper and the gigantic goeduck descend to depths of three or four feet.

This preference for specific habitats is exemplified by many other estuarine species, including the plants. The phytoplankton are the microscopic drifting plants. The transition from salt to fresh water characteristically passes from salt marsh to reed swamp (*Phragmites*) or cattail swamps (*Typha*) or to locations dominated by plants of the sedge family. The dynamic sequence of plant communities also proceeds from eelgrass beds in quiet, submerged shallows to a marsh zone dominated by cordgrass (*Spartina alterniflora*) and succeeded by short cordgrass (*S. patens*) at higher levels where daily submergence is short-term.

A leisurely trip in a small boat around the edges of an estuary can impress you with the variety of animal life. If you are in a contemplative mood, try thinking about the astronomical combinations by which animals that make up the food web get food and survive. Two hundred varieties of fish in a large bay is not unusual; more than two-thousand kinds of organisms may be present in the system. Drs. Andrew McErlean and Catherine Kerby have estimated the possible interactions of these species to number in the trillions.

On your quiet boat trip around an estuary you can see many of the animals that are a part of these interactions in the food web. The consumers are a varied lot. Brant may be seen eating eelgrass in winter. At other seasons, look for moon snails slowly cruising, shrimps sending up clouds of sand, and barnacles and clams opening their shells to filter plankton from the water. More visible are the sanderlings, searching for crustaceans, the yellowlegs, probing in shallow water, and the terns, dowsing for small fish.

Dungeness crab on a substratum of mussels attached to a rock. The majority of these crabs live in at least 300 feet of water. Little crabs change their shells fifteen times or more before they attain the legal sport and commercial sizes of 5¾ and 6¼ inches respectively.

In the broader view, the flora and fauna of estuaries can be classified in groups according to where they occur and what they do. Plant groups include phytoplankton; submerged aquatics, such as eelgrass and many algae; and marsh plants. Animals are represented by bottom dwellers such as crabs, oysters, and clams. These are prey to parasites and larger animals. Drifters are small animals or zooplankton, which include the larval stages of most of the bay animals. The swimmers include resident fish, such as perch and anchovies, spawners, such as shad and herring, and visitors, such as blues, flounder, and cobia. The walkers, creepers, and fliers are the mammals, reptiles, and birds. The wonder of this Noah's Ark of creatures is that all are bound together in an intricate biological relationship that allows the continued existence of each and everyone.

When we become acquainted with a number of different kinds of estuaries, we realize how much the specific assemblages of living things are dependent on varying properties of the environment. The estuarine environment is influenced by climatic conditions, topography, tidal movements, river flows, and salinity stratification. The resulting internal dynamics make estuarine areas of surpassing interest to natural scientists, engineers, recreationists, and fishermen.

All estuaries have one feature in commom—salt water at the seaward end, fresh water at the river end, and the mixture of both in the estuary itself. Seawater has a salt content of 30 to 35 parts of salt per 1,000 parts of water. River water has almost zero salinity. When seawater penetrates the estuary, it flows beneath the lighter river water. This results in a two-layered system of circulation. Water in the upper layer flows toward the sea while bottom water flows upstream. This decreasing salinity gradient from sea to river is influenced by the tides, by wind, by the topography of the estuarine border, and even by the Coriolis effect brought about by the earth's rotation. In large estuaries, this last force creates currents that carry salt water farther on the upstream side than on the downstream side.

Plants and animals adapt themselves in many ways to the circulation and varying salinity of estuarine waters. Plankton species, for example, move upward at night into less saline water and are carried toward the sea. They sink downward in the daytime and are carried by the heavier salt water toward the inner estuary. This type of movement tends to concentrate them in mid-estuary. The blue crab, on the other hand, spawns in salty water near the ocean, and its postlarval stages are transported to less saline waters upstream where food supplies are abundant. After mating, the females move to saltier waters with the aid of downstream currents.

The distribution of sediments, chemical mixtures, water temperature, and boundaries between salinity layers varies enormously among estuaries. Chesapeake Bay receives fresh water from more than 150 tributaries and salt water from the wide entrances of Cape Charles and Cape Henry. The estuaries along the south Atlantic and Gulf shores differ from Chesapeake Bay. These southern water bodies are embayments with restricted inlets from the sea, and they have significant freshwater inflow. Many are protected by barrier islands. The barrier islands restrict tidal currents and permit sediments, of such great value to bottom-inhabiting plants and animals, to accumulate. In contrast, the Laguna Madre, which lies between Padre Island and the Texas coast, is saltier than the Gulf of Mexico. River flow from the mainland is insufficient to compensate for evaporation from the essentially closed bay.

Tidal ranges vary along the different coasts and affect the physical and biological attributes of estuaries. The mean tidal range is 18.2 feet at Eastport, Maine, 4.4 feet at Cape May Harbor, New Jersey, and 1.1 feet at Pensacola Bay entrance, Florida. Monterey Bay, California, has a tidal amplitude of 3.5 feet, and Puget Sound at Elliott, Washington, has a range of 7.6 feet. Circulation forces and flushing rates are better when tidal amplitudes are high. These differences give us a clue to the fascinating variations in plant and animal life and their distribution in important estuaries of the nation.

Few of us have the time or training to study the astonishing details of how estuarine creatures work out their life cycles. And we are not likely to contemplate how most of these animals manage to survive against the physical impediments in the environment and the many predators that have their own life cycles to complete. But some are more than curiosities. Some are delightfully palatable. The oyster is one of these.

My earliest appreciation of oysters began when I was a boy on our eastern Nebraska farm. Each year at Christmastime, my Uncle Elmer, who lived in New Jersey, would send by railroad express a large wooden box packed with bluepoints, which we ate on the half shell. My grandmother also made them into a delicious stew, with hot milk, biscuit crumbs, salt and paper, and gobs of butter. Years later, on my visits to Washington, D.C., and Chesapeake Bay, I graduated from bluepoints to lynnhavens and chincoteagues, caught in the tidal waters of Virginia.

Oysters grow in many areas along our coasts. The Virginia oyster (*Ostrea virginica*) is abundant from Massachusetts southward. Oysters from Apalachicola Bay, Florida, are delectable because of their salty flavor. The coon oyster of the Florida Keys is less desirable for eating. The small Olympic oyster (*O. lurida*) of the Pacific coast is highly edible. I have walked on acres of oysters in Hood Canal west of Seattle when the tide was out.

Oysters live out their lives in estuaries where salt contents vary from about 5 parts per thousand to concentrations approaching those of seawater. They do not live in fresh water and they avoid the sea, with its predatory boring snails, sea stars, and large fish with crushing jaws. Even in estuaries they are plagued by many enemies. Oyster drills—snails that drill through the oyster's shell to reach its flesh —take their toll. So do filtering organisms and small fish that decimate the free-swimming oyster larvae. Oyster populations also are reduced by changes in water turbidity, wide temperature fluctuations, pollutants, and overharversting.

The reproductive practices of oysters vary from species to species. Virginia oysters first develop as males. Some oysters change back and forth from male to female. A female may liberate five billion eggs in a lifetime of ten years. When the eggs are shed into the water they are fertilized by sperm from the males.

The impregnated eggs become larvae with two shells and then glide about for two or three weeks seeking a place to attach themselves. During this stage, they are the size of a pinhead and are known as oyster spat. When there is a spat fall, the young oysters attach themselves to stones, pilings, or other firm places from which they never move again. Within two or three years they reach sexual maturity. In Middle Atlantic areas, they reach marketable size at five years of age. In the warm waters of the Gulf, they grow more rapidly and reach marketable size in less than two years.

In the 1880s, the annual oyster harvest was 15 million bushels. Today, however, their numbers are declining, in Chesapeake Bay and elsewhere. Oystermen tend to resist changes in the harvest level and others believe that the decline in spat fall is a periodic phenomenon. But conservationists remember the problems with oyster beds in Mobile Bay, Long Island Sound, and Delaware Bay and are worried about the future. If we continue to allow the tragic deterioration of water quality, the productivity of oysters and other estuarine creatures could be threatened.

Unlike the sedentary oysters, fish spend their entire lives swimming. Some spend part of their time and others all of their time in estuaries. The American shad leaves the sea, swims through the estuary, and spawns in fresh water. After the eggs hatch, the larvae and young fish live in estuarine waters. As adults, they return to the sea. In contrast, the white perch spends its entire life, from egg to larva to young to adult, in the estuary.

The life cycles of fish are simple in comparison with other animals whose life histories are linked to estuaries. The blue crab, whose life cycle has been described previously, has much more complicated molting, migrating, spawning, and larval stages than any fish. There are crabs and crabs, of course, some of which are tiny and live in oyster shells, while others, such as the spider crab of Alaskan waters,

Oysters growing on cement piers. The colony grows larger since young oysters fasten themselves to the shells of old oysters. The muddy bottom provides no place for attachment, and oysters that do settle there are in danger of being covered with mud by wave and tide action.

measure as much as ten feet between the tips of their large claws.

Hermit crabs are a clan unto themselves. Their soft, unprotected rear ends require that they occupy the empty shells of snails or whelks. As they grow, they have to find ever larger and larger shells. The job of fitting on new suits of armor is a precarious one, since the crab is vulnerable to fish and other predators when it is out of its shell. When it tries on a shell of unsuitable size, it quickly pops back into its original shell and searches for another. At mating time, the male hermit crab remains with a female and drags her shell around until she sheds her skin. Then he deposits sperm within her shell. When the eggs are laid, she attaches them to the appendages of her abdomen, where they remain until the young crabs hatch.

The shrimp is another animal whose life history is linked to estuaries. Like crabs, there are shrimps and there are shrimps. On the Atlantic coast there are mantis shrimps, with forelimbs that resemble the arms of the praying mantis. On the Gulf coast, the shrimps that are caught for food spend their juvenile existence in estuarine waters. On both the Atlantic and the Pacific coasts, pistol shrimps, possessing snapping devices that consist of a trigger at the joint of the big claw, can be found. According to G. E. and N. MacGinitie, a "shot" from the pistol hand paralyzes small fish which the shrimps then grasp and take into their burrows for a meal.

The real burrower among shrimps is the ghost shrimp, found in estuaries from Cape Cod to South Carolina and from Alaska to Baja California. These shrimps spend most of their lives in burrows below the surface of mud, where they dig for the organic matter they eat. Apparently, their muddy homes must have exactly the right consistency, kind of food, and feel to their bodies. Otherwise, they die.

Another creature with a strange life cycle is the northern sea horse. It lives among weeds and eelgrass in sheltered bay waters south of Cape Cod. A relative, the dwarf sea horse, lives in the turtle grass along the west coast of Florida. These bony, spiny, armored fish cling vertically with prehensile tails to grass stems while they feed by sucking small animals into their round mouths. They can swim slowly by vibrating their earlike pectoral fins and their dorsal fin. But mostly they are content to remain motionless and unseen among weeds and grass stems.

The female sea horse deposits her eggs on the abdomen of the male, which incubates them in a pouch formed by his pelvic fins.

When the young are born, they swim horizontally, even though they have a prehensile tail. Eventually, they assume the vertical position and cling to weeds and grass while they wait for crustaceans and other small creatures in their underwater grass jungles.

Another group of unique fish that live in estuaries, bays, and in the open sea are the flatfish. There are 500 species of flatfish found in the sea: seventeen kinds are recognized along the Oregon shore, at least 20 species live along the Gulf shore, and numerous species frequent the waters south of Cape Hatteras. Flatfish have many names, including flounders and turbots. Some of the more exotic names include the ocellated flounder, broad flounder, southern flounder, fringed flounder, spotted whiff, bay whiff, fringed sole, lined sole, and hogchoker.

This reminds me of a discussion I heard one day at Yaquina Bay on the Oregon coast. Some boatmen were examining a freshly caught flatfish and trying to decide if it was a flounder, a fluke, a sole, or a dab. One thought it was a "rock sole," which is not a true sole at all. Since the fish was only seven inches long, they solved the problem by throwing the "poor soul" back into the bay.

There are "left-eyed" and "right-eyed" flatfish. The left-eyed species have their eyes and coloration on the left side, which means that the fish rests on its right side next to the bottom. This flattened profile begins when the fish larvae are less than an inch long. One eye and the mouth migrate to one side as the gills and internal organs also become rearranged. The blind or bottom side of the fish is usually white, while the top side is colored and is capable of changing hue to correspond with the color of the bottom sand or mud on which the fish rests. Some flatfish show beautiful mottled patterns.

The sand sole (*Psettichthys melanostictus*) of Puget Sound waters, and other soles and flounders that occur in shallow waters at low tide, may be stirred up by wading. Even then, they are difficult to spear or catch by holding your foot over them. They make a quick dash and then wriggle into the sand or mud so that only their eyes and mouth are visible. Theirs is a peculiar life, one spent lying on their right or left sides. Fortunately, the worms, crustaceans, and other small animals they eat are right beneath them in the sand or mud.

As in all biotic entities, the estuarine life system begins with plants, or producers, and their photosynthetic capabilities of storing

Crab pots made of rope and netting are baited and lowered into bays. Weights hold the pot on the bottom, and a buoy floats on the surface to mark its location.

energy. They have little or no control over the consumers, the links in the chain that extends from zooplankton to fish, birds, reptiles, and mammals.

The consumers are a varied lot. From the small to the large, they vanquish one another in the continuing feast that is part of the good life for each animal. Their feeding habits are widely diversified. Some feed on plankton only, while others eat detritus. Among them are strict herbivores and strict carnivores. A few eat only shrimp or are limited to one type of food only. The higher carnivores, especially some of the fish, eat two, three, or even more than four kinds of food and are not confined to one trophic level.

This one-way flow of energy from the sun through green plants to animals results in food chains that become part of food webs. When a brant or other seabird eats eelgrass it also eats innumerable organisms attached to the eelgrass, thus affecting more than one food chain. Destruction of one link in a chain alters the food web and results in a dead end for certain organisms. When disease nearly eliminated the eelgrass some years ago on the Atlantic coast, it almost spelled the end of brant populations.

Food chains do not always begin with living material. Much decaying plant detritus is eaten by estuarine animals. Cordgrass (*Spartina alterniflora*) detritus is a protein-rich plant material that is food for worms and other bottom-dwelling species. On the bottom, bacteria and diatoms grow in numbers that reach millions per quart of water. They are filtered by multitudes of copepods, larvae of various animals, and shrimplike species.

These small animals, in turn, are ingested by larger animals, including annelid worms, crustaceans, mollusks, and the young of various fish. Much of the material consumed is returned to the estuary by evacuation and by decay of the animals themselves when they die. This is pumped to the surface of the bottom mud or sand by burrowing organisms. From this we see that an estuary is a sink where essential elements, such as phosphorus and nitrogen, are recycled again and again to the benefit of living things.

Young snowy egrets climb out of their nests and walk on tree limbs long before they are ready to assume a life of fishing in the marshes.

8

Salt Marsh Mysteries

THE FIRST MARSH I EVER SAW was not a salt marsh at all. But it was filled with mystery and beautiful creatures. Its reeds and rushes murmured in the wind with the sweet softness of a Brahms symphony. The marsh was the remnant of an oxbow lake, formed when the Missouri River changed its course and left behind the old channel to become a lake and, later, a freshwater marsh.

I first saw the marsh when I was about ten, when my parents and some neighbors went there on a fishing expedition. The fishing was fun. The water of the diminishing lake was rampant with crappies, sunfish, bullheads, buffalofish, and turtles and so shallow that we could catch the fish with our hands. But I found it more interesting to wander off into the fringing reeds and bullrushes and cattails where redwing blackbirds nested by the hundreds, muskrats had cut trails through the tangled stems, and raccoons had left their slender finger marks beside crayfish funnels they had explored for food. Thus was I introduced to one of the most proliferous and valuable plant communities on earth.

There were other fascinating creatures in and around this marsh. Snails clung to the waterweeds near the surface of the spongy mat beneath my feet. Grasshoppers moved behind the stems of cattails at my approach, making themselves invisible, as do squirrels when a hunter approaches them in an oak woods. I discovered a duck's nest with broken eggshells—possibly the remnants of a skunk's dinner. When I emerged on a mud flat which had not yet been conquered by the marsh vegetation, the place was littered with plovers, sand-pipers, avocets, and other birds that I could not name at that tender age.

A bald eagle, the first one I had ever seen, floated in the sky above, but I did not then connect its presence with the food chain of marshes, which extends from bacteria, copepods, mosquitoes, min-nows, larger fish, snakes, small birds, mice, and frogs to foxes, minks, great blue herons, and eagles. Even my parents and fishermen neighbors never contemplated the biological significance of this lake and marsh, though it was an ecological community very different from the familiar forests and fields of their farms.

As all marshes are, this one was a productive area. Over a period of some thirty years, the Missouri River occasionally inundated the place at flood stage, adding new fish to the lake and soil and muck to the marsh. Rain and small creeks also carried nutrients down from adjoining river bluffs.

Eventually the lake was drained to become farmland. The soil soon demonstrated its productivity by growing corn so tall the farmer who owned the place facetiously remarked that he had to stand in his wagon to pick the highest ears. Then he decided to plant tomatoes among the corn and fertilize the field with nitrogen to pro-duce the greatest tomato crop ever seen. By that time I had done graduate work in plant physiology and advised him not to use fertil-izer on so rich a soil, but he disregarded my advice and produced a veritable jungle of tomato vines that climbed ten feet high on the cornstalks. In fact, the whole field went to vines. He never har-vested even a bushel of fruit from each acre of tomatoes he had planted.

This experience sparked my lasting interest in wetlands of all kinds, including bogs, swamps, and marshes. It even led to my doc-toral dissertation on the ecological and physiological adaptations of the swamp or tussock sedge (*Carex stricta*), which grows in Wiscon-sin, Connecticut, Hungary, and elsewhere. Since then, other inves-tigators have made detailed studies, particularly of halophytes (plants that endure salt water) and of coastal marsh productivity and its importance in the whole scheme of plant, animal, and human life.

Over the years I have visited many marshes on all our coasts. In the salt marshes there are always many things to see and feel. I re-member meeting a man in the marsh at Barnstable on Cape Cod. He was not looking at the gulls circling and crying overhead. In-

Black-crowned night heron on the nest. This bird, which utters a "quawk" at night, occurs inland near marshes and swamps and along the coasts of North America and of other continents.

stead, he mentioned the subtle tints and shades of the emerald and gold cordgrass and the gentle lines where the sea, the marsh, and the shore merged in artistic curves and lines. He remarked that the profound beauty of a marsh reflected in the sunlight can cause pain of a pleasurable kind. I had to concur.

On other occasions, I have gone on pleasant excursions in the Virginia marshes with fishermen who were using mummichogs (killifish) as bait to catch summer flounder. The flounder spawn in the marsh creeks and feed in the channels where the water goes back and forth with the tides. In summer, the fishermen bait funnel traps with bacon to catch silversides, black drum, bluefish, small crabs, and even eels. These waters are also the home of the delightful sheepshead minnow, with its breeding hues of salmon, metallic green, and blue.

On these trips I must confess that my interest frequently wandered away from fish. I found myself watching for those chickenlike birds, the clapper rails, that are symbolic of water birds in Atlantic marshes. Their presence indicates the presence of other creatures, including small fish, crustaceans, insect, and fiddler crabs. The laughing gulls are also here in summer. They nest on barrier islands and fish in the open waters of marshes as well as in deeper waters, where bluefish drive baitfish to the surface. Other gulls also frequent the marshes. In winter, when pickings are lean along the storm-swept sandy beaches, the ring-billed gulls are numerous in the marshes and meadows of both the Atlantic and Pacific coasts.

Some knowledge of geology and ecology is helpful in understanding how marshes are formed. Sinking of the seacoast can set the foundation for a marsh. Barrier islands and sandspits can reduce the tidal action and cause sediments to settle in lagoons between the islands and the sea. As the waters become shallow, halophytes invade. Through plant succession, one species follows another, each replacing the preceding one. The result is a varied group of plant communities that lend variety and mystery to these quiet, shallow, and productive tidewaters.

Some marshes border mud flats that are covered with seawater at high tide. These are exposed at low tide and, since they are made productive by small plants, mainly diatoms, and are abundantly occupied by worms, crustaceans, and other small animals, they are especially attractive to shorebirds. The marsh itself, with its tall grasses, is an ecotone, or transition area, between the marine and the terrestrial habitat.

As the roots of cordgrass and other herbaceous plants decay and trap sediments brought in from the sea, peat is built up and the marsh ground level rises above the tide level to become a freshwater marsh, a meadow, or even a forested area. If the sea level rises, the marsh expands and increases its intricate pattern of tidal creeks and channels.

Many of the marshes along the New England coast are now drowned areas that once were occupied by freshwater marshes and ponds. In some tidewater areas in Virginia, marsh grasses and debris have filled entire coves that have since been transformed into fertile farmland. Along the California coast, marshes form fringes at the edges of various harbors. These vegetation zones are usually narrow and correspond mostly with the area exposed at mean high tide. Above the highest tide levels, the vegetation changes to the terrestrial plant species adapted to salt spray and seepage of fresh water from the uplands.

If the sea level is stable, a series of communities develops in zones or patches in accord with the depth of the salt water and the degree of inundation by the tides. These communities can be distinguished by the kinds of plants dominating the various zones. Below mean tide level, saltwater cordgrass, small cordgrass, and the salt marsh bullrush are the dominant species. The higher levels of the marsh, covered only by spring or storm tides, form a salt meadow where herbs such as seaside goldenrod, sea lavender, and seaside aster flourish.

The presence of fresh water in the marsh is indicated by the common cattail (*Typha latifolia*) and the narrow-leaved cattail (*T. augustifolia*). Cattail communities are thick with buried rootstocks and tubers. Here in these drier portions of the marsh, muskrats feed and build their dome-shaped houses, weasels prowl for meadow mice, and minks and blacksnakes search for mice and frogs.

Shrub and tree communities border many marshes. Above the continuously wet zone, the marsh bayberry and beach plum grow in company with cardinal flowers, arrowheads, sweet gale, and buttonbush. Mallows with big pink flowers add variety in the transition zone between freshwater and dry-land species, which vary from pitch pine to red maple, oak, and hickory.

The wonders of marsh and meadow plants are not all aboveground or even beneath the water's surface. Few of us are aware of the internal adaptations of the plants themselves as we tread their spongy substrata or explore marshy creeks and tidal channels with boat or canoe. The roots of the grasses, for example, are so abundant that a square yard of marsh bottom may contain living fibers and branches that would extend for miles if placed end to end. Also unknown are the internal adaptations that enable these plants to persist in an environment that subjects them to salt water from the sea, fresh water from inland streams, and alternate inundation by tides and exposure to drying winds when summer drought tends to desiccate their living substance.

I discovered some of the mysteries of marsh plants when I studied the tussock sedge for several seasons. This plant intrigued me because of its habit of producing tufts of roots from one to two feet high, surmounted by grasslike leaves that grew another two or three feet high from the crest of the root tuft, thus forming a tussock sometimes half as tall as a man.

The anomalous feature of the plant was that water usually stood at the level of the leaf bases in spring, while in autumn the water table subsided to depths of one to three feet below the ground level, leaving the roots of the tussock trunk exposed to sunlight and air.

The question was, how could the plant withstand such violent fluctuations between extreme wetness and equally extreme dryness?

The answer lay in the internal characteristics of the plant itself. During that portion of the season when the water was well above the surface of the ground, the leaves and stems at the top of the pedestal of fibrous and wirelike roots were seldom, if ever, submerged. The roots and rhizomes possessed air channels, which supplied oxygen to the submerged plant parts. The wirelike roots, which were exposed to the sun in dry seasons, contained several layers of sclerenchyma, or lignified cells, which prevented evaporation from the roots. In addition, the leaves possessed "motor" cells, which collapsed and allowed the leaf blades to fold, thus reducing their evaporating surfaces during hot dry weather. So here was a plant that was adapted both to marshy conditions and to summer drought.

The salt-marsh plants along the seashore have other characteristics that adapt them to their watery world. The large cordgrass (*Spartina alterniflora*) along the Atlantic coast can germinate seeds in salt water. The grass spreads by rootstalks, and the coarse green stems grow to a height of six to ten feet, anchored by intertwined roots on muddy bottoms. The plant thrives when covered by tides for half of each day. Some of these stands are like small forests. If the sea level does not change, they ultimately destroy themselves by depositing so much debris from leaves and roots that the marsh level rises.

Then the salt-meadow grass (*Spartina patens*), which gets a bath only during high tides, takes over. This grass, with wiry stems and slender leaves, is the dominant grass in many shallow marshes. Its leaves seldom grow more than two feet tall and become matted together by the wind so they resemble cowlicks. Salt-meadow grass is one of the salt-hay species of grasses which formerly were used extensively to feed cattle, thatch roofs, and pack fragile articles in wooden boxes for shipment.

The large cordgrass also used to be mowed for hay. If the marsh was shallow, farmers pastured their cattle there when the grasses were still green. Cut hay was sometimes piled in stacks, held in place above the marsh wetness, but, occasionally, high tides washed away the cut hay before it was stacked. In New Jersey, salt hay is still baled and shipped for mulch in gardens and for cover on strawberry beds. Dams are constructed across creeks and tidal channels to prevent inundation of the marsh, but sluice gates are opened at the proper time to allow the tides to fertilize the hay meadows.

Along the high edges of marshes, black grass (*Juncus gerardi*), which is a rush, not a grass, is a common plant. It produces nearly black flowers. Spike grass (*Distichlis spicata*) also grows in these drier sites. It becomes straw-colored in winter in contrast with the gray of the cordgrasses.

Cordgrass (*Spartina foliosa*) has been introduced to the California coast. Like the tussock sedges I studied, it has various adaptations for growth in watery environments. The leaves and roots have hollow passages so that air can move through the plant even when it is partially submerged by the tides. It also has special glands to excrete salt, which forms little white crystals on the leaves.

Other grasses and salt-enduring species, especially along the Atlantic shore, include salt reed grass (*Spartina cynosuroides*), whose brown to purple spikes reach heights of 3 to 9 feet in brackish water; the reed *Phragmites communis*, which grows to heights of 13 feet in hedgelike rows along ditches; the soft rush (*Juncus effusus* var. *solutus*), which grows up to 6 feet in salt marshes; and the salt marsh bullrush *(Scirpus maritimus)*, whose stems, unlike those of the grasses, are distinctly three-sided and reach 3 feet in height.

One of the notable marine plants found in shallow salt water along the Gulf coast is turtle grass (*Thalassia testudinum*). In appearance it resembles a grass, but, along with some strange relatives, such as tape grass (*Vallisneria americana*), also called eelgrass and water celery, and waterweed or ditch moss (*Elodea*), commonly grown in fish aquaria, it belongs to the frogbit family. Unlike some of its relatives, turtle grass is pollinated beneath the surface of the water so it needs neither insects nor wind to help fertilize its female flowers. It grows extensively in saltwater marine meadows from Florida to Texas and in South America. Turtle grass furnishes food and protection for large numbers of marine animals because of its vigorous capacity to produce leaves from thick creeping scaly rhizomes that cling to the bottom in bays, marshes, and beach drift.

The plant enthusiast can find many other species of halophytes in marshes near the sea. The pickleweed (*Allenrolfea occidentalis*) is a fleshy plant with segmented stems that have no leaves. One of its adaptations is the ability to store fresh water in its tissues by reverse osmosis. The plant itself is green and succulent in summer and an attractive scarlet in autumn. When dry, its woody twiglets help stabilize the banks of sloughs and marshes. Crabs like to burrow among the decaying root debris of pickleweed. This aerates the sediments where other animals also live.

The woody glasswort (*Salicornia bigelovii*) is a perennial that seldom exceeds four inches in height. Like the pickleweed, it turns red in fall. The jointed glasswort (*S. europaea*) is an annual halophyte that grows upright like a small tree. Ducks relish the seeds of these plants, and geese eat the stems.

In summer, the marshes are brightened by many plants with showy flowers. Among these are the rose mallow (*Hibiscus moscheutos*), seaside gerardia (*Gerardia maritima*), and the sea lavender (*Limonium carolinianum*), with its leathery leaves. The flowers form a purple fringe along marsh edges in autumn. Along with the grasses, bullrushes, and the marsh rosemary, the flowering plants present an extravagant panorama of color and beauty that extends from sea level to highland grassy meadows and forests, flourishing beyond the reach of the ocean.

As any visitor to a salt marsh knows, the summer marsh is not without insects. The fireflies flashing their mating lights add a touch of radiance to the expanse of grass-covered water. The salt marsh mosquitoes (*Aedes solicitans*) lay their eggs in tide pools and puddles. Mosquitoes become so numerous at times that they resemble clouds of gnats. Their biting can drive you into a mental and physical frenzy if you are not properly clothed and protected by insect repellents.

Even more vicious bloodsuckers are the female greenhead flies (*Tabanus nigrovittatus*) that rise from the marshes and even invade beaches for a week or two in midsummer. They lay their eggs on stalks of cordgrass. But they redeem some of their nuisance tactics when their eggs hatch and the wormlike young feed on other insects and snails in the marsh. These flies also encourage the presence of more birds in the marsh—especially the seaside sparrow, which nests among the cordgrass plants and includes greenheads and other insects in its diet. Likewise, the larvae of mosquitoes are food for tiny fish which, in turn, are food for larger fish and for water birds. Certain insects also pollinate the flowers of marsh herbs and shrubs. Thus, in their own ways, insects spell success or failure for many living things and are an important part of the marsh ecology.

Unless we look closely, we do not see how insects and other animals are bound together in a system that affects the principles and processes of the marsh. On a canoe excursion we may see the salt marsh caterpillars (*Estigmene acrea*) nibbling on grass stalks. But we do not see the beetle grubs chewing on the grass roots unless we dig into the spongy mat that covers the bottom of the marsh. This mat retains moisture for many creatures, from desiccating bacteria, minute crustaceans, snails, and fiddler crabs to the whimbrels, raccoons, and other animals that feed on them.

Each vegetation community in the marsh has its own set of insects. In the cattail marsh there is more than the world of Typhaceae, with its stems that stand side by side and its stiff upturned leaves. The cylindrical brown heads that send out their fluffy seeds on the winter winds are home to an insect called the cattail moth (*Lymnaecia phragmatella*). It is a widely distributed insect, since the cattails on which it depends entirely are circumboreal. I have found these tiny creatures in the cattail swamps of Oregon, in the prairie ponds of Nebraska, and in the salt marshes of Virginia. The adults lay their eggs only in the pistillate catkins of the cattail and there the larvae remain, eating only the seeds and seed heads until they become adults a year later.

The cattail also plays host to other visitors, including the cattail miner moth (*Arzama obliqua*), which dwells inside the stalk of the cattail and never interferes or communicates with the grasshoppers that cling to the outsides of the stems or with the larvae of the snout beetle, which live in the rootstalk in the muddy bottom of the marsh. But the dragonfly, sitting on its favorite cattail leaf between flights, undoubtedly captures the adults of the miner moths as well as innumerable mosquitoes when they are in season.

When the redwing blackbird flocks break up in spring, many individuals scatter to the marshes to build their nests among the cattails, sedges, and grasses. Grass roots, mud, the lint of milkweeds, and the leaves of slender rushes and cattails are used to tie the nests to the stalks of tall marsh plants. And the marsh supplies an abundance of food for the fledglings, including beetles, spiders, moths, caterpillars, and grasshoppers.

Grasshoppers are intriguing insects with a number of interesting habits. Many are inclined to roost at night on plants they do not eat. All are experts at hiding by moving behind grass stems or leaves. *Leptysma* grasshoppers are very slender and have faces that slant forward. In the vast marshlands, their form and color enable them to vanish like magic.

One of these, the cattail toothpick grasshopper (*Leptysma marginicollis*), an old friend I used to know in the swamps on our Nebraska farm, lives in marshes from Maryland to Florida to Texas and Nebraska. It is slender, brownish or green, often suffused with pink, marked with a yellowish stripe on the side, green beneath, and embellished with bluish hind tibia. It is adept at hiding, quick in flight, and hard to catch, especially in the vast entanglements of cordgrass in the eastern salt marshes.

Other insects that thrive in marshes are the plant hoppers, which suck the sap from grass leaves and stalks. Numerous flies feed on plant secretions and rear their larvae in decomposed plant parts. Thousands of springtails and beetles live in densely vegetated areas. Eye gnats of the genus *Hippelates* swarm about one's eyes and moist wounds and are a distinct nuisance.

Little is known of the place that many insects occupy in marsh ecology. The insects have predators, chief of which are dragonflies, spiders, bugs that hunt herbivorous insects, and parasitic wasps. Ants that live on the landward sides of marshes also hunt insects. The variety is large since some 400 kinds of insects are known to inhabit marshes. As usual, they are links in various food chains that start with insects that eat bacteria, organic plant debris, and the leaves and roots of living plants. Spiders and insects are, in turn, eaten by birds.

Maybe the insects perform another function in marshes, especially the bloodthirsty ones that keep out of the marshes at least some humans who disturb wildlife. There are 67 species of mosquitoes in Florida. Head nets, plastic bags around the feet, plastic gloves, and chemical sprays reduce some of their attacks on people. Sand flies are even harder to repel, but diethyl toluamide seems to help. The deerfly is another infamous creature. But none of these insects, or anything else, seems to repel the rubbish dumpers and the exploiters who believe that the best use of a salt marsh is to bulldoze it for occupation by houses, factories, roads, and the other works of man.

Nature is not benevolent in a salt marsh. The animals that live there must contend with the elements and with living enemies, as do all creatures that are links in the chain of life. The alewives (*Alosa pseudoharengus*), which migrate through the marshes to spawn in fresh water, attract gulls, blue jays, and grackles. The fiddler crabs that scurry through the grass thickets feed on minute plants but, in turn, are food for whimbrels, gulls, and raccoons.

The insects that live both above and below the ground are prey to moles that tunnel in the drier soil at the edge of the marsh and to shrews that can eat their own weight of animal flesh each day. Even the marsh snails must climb out of the water to breathe since they have lungs rather than gills. If they escape the fish and other submarine predators, they become victims of the clapper rails and other marsh-inhabiting birds.

The lives of marsh animals, like the lives of animals of other ecosystems, are biologically determined, at least insofar as their needs for food and reproduction are concerned. Their survival as individual species also depends on shelter, their genetically established length of life, and behavior, which adapts them to a watery, shady, salty environment. The salt-marsh harvest mice that nibble on pickleweed in San Francisco Bay's marshes, now on the endangered species list, exist in their marginal habitat because mouse power comes from rapid reproduction. The diamondback terrapin, which lives out its life in eastern coastal marshes and catches fish, crabs, and snails, has no need for a monthly crop of youngsters. It lives for years, and its destiny is determined less by natural enemies than it is by the destruction of its habitat, caused by the replacement of vegetation by brushland and then forest or by man's dredging and pollution of its environment.

A more adaptable animal, the raccoon, which finds abundant food in the marsh, simply migrates to the river, the forest, or even the city if its marshy domain is destroyed. I well remember one particular raccoon that used to wait for the traffic light to turn green on Sandy Boulevard in Portland, Oregon, before crossing 82nd Avenue on its way to the tree-covered bluffs behind Children's Hospital. Other raccoons also became city dwellers when newly constructed motels near the Portland International Airport destroyed part of their swampy habitat. The mink and the otter, on the other hand, do not move to the city when the marsh is destroyed; they migrate to other watery habitats. And the rice rat of southern marshes is simply out of luck if its habitat is destroyed.

The salt marsh, then, is a place where one can observe animals that range from lobsters in their soft shells to shrimp on muddy bottoms and snails that close horny operculums against evaporation

Great blue heron in tree nest. Even in northern climates, eggs are laid and the young are hatched before leaves appear on deciduous trees. The adults sometimes fly many miles to obtain fish for their young.

when the tide is out. There are eels that traverse the marshes to freshwater streams. And there are ribbed mussels (*Modiolus demissus*) that anchor themselves to the bases of grasses and form thick beds in the spongy bottom. All these live in a world of moisture, a great sponge, that acts as a buffer between land and sea and protects them from drought, storm waves, and flood tides.

The birds, those most mobile of animals, nearly always have representatives in the marshes. Few are permanent residents. Most are migrant species that come there because of the marsh's seasonal attractions. In northern marshes, where the water freezes in winter, owls are common visitors since mice, shrews, and muskrats still are active. In the southern marshes of Florida and the Gulf coast, wading birds may be seen the year round.

Great blue herons, the largest wading birds on any of our coasts, are among the most spectacular of the marsh birds. Their nesting habits make them adaptable to many environments so long as there is water with fish, frogs, small rodents, and other food items within a few miles of their rookeries. I have seen their nests built 200 feet above the ground in Douglas fir trees on Vancouver Island, British Columbia. They nest in ponderosa pine trees along the Tumalo River in Oregon. They use cottonwood trees for their rookeries in the sandhills of Nebraska where the nearest fishing is eight to fifteen miles away on the Dismal River. They use mangrove trees in southern Florida. These giant birds also have the patience of Job while waiting for fish to swim within reach of their rapierlike bills.

On eastern and southern shores, the marsh-inhabiting birds include the great blue heron, the little blue heron, the Louisiana heron, the great egret, and the snowy egret. Less frequently seen are the common snipe, the clapper rail, and the black rail. The last-named is an especially shy, sparrow-sized bird that disappears into the sedges and the cordgrass jungles at the slightest disturbance. The Virginia rail shows its presence by flying up and then disappearing. The least bittern, camouflaged by its striped brown plumage and vertical stance among the grasses, is common in cattail marshes. It starts up with a squawk if approached closely.

Winter residents and migratory birds in southern marshes include Virginia rails, herons, curlews, sandpipers, plovers, ducks, and geese. The variety of habitats that occur in marshes determines to some extent where different species do their feeding. Canada geese prefer the borders, channels, and ditches. Birds that probe for worms, mollusks, and small crustaceans roam the mud flats when the tide is out. The deep quacking of black ducks far out over the marsh sometimes indicates the presence of large ponds and open water.

Along with the grasses, all the birds, mammals, fish, and lesser animals contribute to the fertility of the marsh. The influence of this ecosystem extends beyond its own limits to the land and to the sea. At high tide, silt is deposited in the grass thickets, creating a substratum where burrowing animals can live. The outgoing tide carries organic matter from decayed vegetation and the droppings of birds and other animals to the bays and estuaries. Here, and in the sea, the gift of the marshes enriches the waters and provides food for planktonic creatures and the larvae of scallops, fish, oysters, and clams. We can only conclude that the salt marsh is a place of complexity, mystery, and abundance which we have hardly started to measure and which we do not completely understand.

Author examining a huge tree stump brought to shore by an ocean storm. The year before the storm this was a sandy beach. Even the cobblestones were not there.

9

The Exotic Gulf Coast

THE GULF COAST, FROM FLORIDA TO TEXAS, with its wealth of natural beauty, its rich flora and fauna, and its unique environments, has attractions for almost any visitor. The white sands, palm trees, coral reefs, marshes, swamps, barrier islands, and tropical seas give the coast an aura of enchantment. The environment is dynamic and ever-changing.

Sunny days on beautiful beaches can be succeeded by ecological turbulence, when violent hurricanes destroy islands and shoreline habitats of birds and mammals, even people. But nature restores itself and once more the fiddler crabs bestir themselves among the mangroves, the grasses catch hold of the sand and rebuild the dunes, and the wading birds explore the tide line while the least terns lay their eggs at the edges of the dunes.

There are many things to see on sandy shores. After a storm the beach is a cornucopia of shells. Among these are the large conches, pen shells, and moon shells, with spiral patterns and sand rings that protect their eggs. If you are lucky, you may find a white wentletrap with its spiral staircase shell. The egg sac necklaces of the whelk and the purselike egg containers of the skate are other shell treasures. The jetsam and flotsam washed in by the tides also yield driftwood, bottles, floats, and an endless variety of animal life, from jellyfish to stony corals, starfish, and sand dollars.

There are no mountains in the landscapes of the Gulf coast. The variety lies in the sandy beaches, dunes, salt marshes, and forested areas, with many trees festooned with Spanish moss. Along the shores are endless miles of swamps and tranquil rivers that flow through cypress forests and bayous to the Gulf of Mexico. Not the least of these rivers is the mighty Mississippi, which drains a large share of the continent.

The Gulf coast is low country, a land with much water. The houses of the swamp dwellers rest on stilts that prevent flooding when hurricane winds and sea tides raise the water level to living-room heights. New Orleans is so waterlogged that huge pumps drain 30 million gallons of water from beneath the city even when the weather is dry. The French Quarter is only 14 feet above sea level. As Bill Crider, a news reporter, recently described it, this is "high enough to make a swamp dweller's ears pop."

The people of the bayou country and the alluvial swamps are as adapted to life there as are the graceful wading birds, the muskrats, the fish, and the shrimps that come up from the sea. Many of these animals derive their sustenance from waters made nutrient rich by soil from Ohio, Iowa, Missouri, Louisiana, and other states where the overflow comes in rivers from the north.

Some swamp people live lives of isolation. I remember the night when, just as I was falling asleep on my camp cot in a Mississippi swamp, the end of my bed sagged a few inches and I looked up, startled, to find a man sitting by my feet. He was a swamp dweller, and he had come to talk about the world beyond his own.

We chatted for an hour, and he admired the canoe my partner and I had paddled through channels in the swamp. Before he walked away in the darkness, I told him we would enjoy having him eat breakfast with us in the morning. But I had some doubt about finding anything that might serve as fuel for a fire over which we could cook our ham and eggs and heat our coffee. The pelting rain of the previous day had dampened every blade of dry grass and piece of wood on the soggy ground. He said nothing.

At first daylight he appeared with an iron gadget about a foot long that looked like a fisherman's gigantic treble hook. A rope was attached to an iron ring at the end opposite the hooks and he swirled the thing around like a lasso, tossed it up into some dead trees, and pulled down dry limbs for the breakfast fire. I realized I still had much to learn about camping and living off the country in the strange land of the bayous. I also learned how men, as well as wild creatures, have adapted themselves to life in the lowland environment of the Gulf coast.

The marshes, sand beaches, shell beaches, and mud flats of the Gulf coast are the result of recent historical geologic change. Modifications also have resulted from erosion, movement of sediments by sea currents along the shore, tidal action, and storm waves. Although normal tidal action is approximately 1.5 feet, hurricanes can raise tides of 15 feet or more.

Geologic change is still in progress. The Louisiana coastal marsh is subsiding at a rate of about four tenths of a foot per century. Since the average level of the marsh is only one or two feet above sea level, the more the land sinks, the more the marsh becomes vulnerable to deterioration, attack by waves, and invasion by seawater.

Sea level, at the time of the Pleistocene epoch, was about 450 feet below its present height. The Gulf shoreline extended many miles to the south. As the continental glaciers melted and the level of the sea rose, rivers began to fill their inundated valleys with sediment and the sea moved inland. The Mississippi and other rivers deposited sediments that were then transported along the shore by sea currents. In the last four thousand years, the Mississippi has shifted its course eastward and westward. Within the last fifty years, the Atchafalaya River, one of the distributaries of the Mississippi, has increased in size.

The Atchafalaya now is an enormous alluvial swamp, fascinating because it is wet, wild, and home for a diverse flora and fauna. The Army Corps of Engineers wants to dredge a large channel through the basin to flush millions of tons of sediment from the Mississippi River into the Gulf of Mexico. If they accomplish their purpose on a grand scale, the fragile ecosystem that has been built in the basin through centuries of natural evolution may destroy part of the basin and much of its wildlife.

Much of New Orleans now rests on land 6 feet below sea level. The levees stand so high that men on passing ships look down on dwellings and on people below on the city streets. If the overflow of the Mississippi is diverted through the Atchafalaya basin, its sediment will be transported to Atchafalaya Bay and to the Gulf. In time, this will form new mud flats and marshes and the coastline will continue its never-ending change.

A hurricane is one of the greatest catastrophes that can come to the Gulf coast. Even if no hurricane has developed, strong winds may whirl in unexpectedly from the sea. On May 31, 1975, my wife and I drove to Panama City, Florida, creeping through what seemed like miles of vacationing students and tourists. It was a sunny day, and all was peaceful. People carpeted the beach and the highway in an almost impenetrable mass. Then a near gale came off the Gulf of Mexico, and high waves began running up the silvery sands that stretched westward toward Pensacola. It was difficult to stand upright against the wind. We tried catching ghost crabs, but they were too nimble, even in the wind, and invariably disappeared into the grass clumps on the miniature dunes.

In September of that year Hurricane Eloise struck Panama City. The resort was unprepared for the 130-mile-per-hour winds. Predictions of Eloise's direction were off course, even though the warnings had included a 325-mile strip from Louisiana to Apalachicola, Florida. A hurricane had not struck Panama City since 1900, and people did not believe it would happen again.

Returning homeowners were shocked as they looked at caved-in houses, uprooted palm trees, and overturned concrete shelters and tables in the wayside park. Beach cottages were flattened. The U.S. highway was closed and boats were sunk in the community marina. And Hurricane Eloise was not as bad as Hurricane Camille, which struck Biloxi, Mississippi, in 1969 and did even more damage.

No one knows for certain when and where a hurricane will strike. In late August 1977, Hurricane Anita slammed into the northeast Mexican coast, washing out roads and uprooting trees. The residents of Soto la Marina and La Pesca, on the edge of the Laguna Madre, were evacuated by army personnel just ahead of torrential rains and howling winds. The erratic nature of Anita caused thousands of residents of coastal towns in Texas and Louisiana to evacuate. Cameron Parish officials closed schools. Storm shutters were put up in New Orleans. Relief stations were established while workers began checking more than 100 miles of levees and flood gates. People who lived more than 5 feet below sea level were told to be ready to move out.

In September 1974, the killer hurricane Carmen smashed into the Louisiana Gulf coast and more than 100,000 persons fled their homes ahead of winds with gusts up to 150 miles per hour. People still remembered the devastation caused by Hurricane Camille in 1969. They also remembered Hurricane Betsy and other great storms in the past that had taken many lives and caused hundreds of millions of dollars' worth of damage.

Hurricanes are not limited to the Gulf coast. These violent

storms also swirl up the Atlantic coast from Florida to New England. In 1635 in early August, a tide 14 feet higher than normal was reported at Narragansett in Rhode Island. On August 16, 1899, a great storm named San Ciriaco struck near Cape Hatteras. Even Columbus, on his voyage in 1494, apparently encountered a hurricane.

I have never been in a full-blown hurricane. But in 1925 I rode the Flagler Railroad from Key West to Miami when a tropical storm was blowing. The waves were running high as we crossed the bridges between the Keys and seawater was splashing against the Pullman windows. Trees were being whipped wildly, tossing their leaves into the gale. No great damage was done to the buildings.

Years later, I experienced at first hand the eerie effects of a high wind on the plains of Colorado. A chinook wind that reached 110 miles per hour blew away buildings, tore roofs off houses, and overturned automobiles while three of us were hunting ducks on Cobb Lake northeast of Fort Collins. Some 50,000 ducks were rafted there on the ice and the wind lifted them like a living carpet that tore apart in the gale. Thousands of birds fluttered vainly in an attempt to make headway against the wind.

That was the day the ducks flew backward. And it was the day we crawled on the ice, wind-blasted by drifting sand. For the next two days, we were incapacitated with what resembled severe arthritic pains. The doctors who treated us believed our pains came from static electricity generated by the dry sand spray. Our experience with the wind was like that of a hurricane over seaside dunes, but without the torrential rains that usually accompany hurricanes.

Coastal hurricanes frequently produce giant tidal waves that sweep across beaches, towns, marshes, and forests. Old inlets between islands are opened or new ones are formed. They obliterate some of man's tinkering with the seashore. Occasionally they make us cautious about building houses too close to the sea.

Hurricanes change landscapes, move islands, dig new inlets for estuaries, and destroy vegetation and wildlife. But their devastation promotes the succession of new floras and combinations of environments. In essence, they provide for the continuation of species of plants and animals that would otherwise tend to disappear in stabilized living communities. Thus, they preserve the genetic reserve of life forms.

Hurricane formation is still not exactly understood. These giant storms, intense cells of low pressure with a central eye surrounded by a circular wind system, are generated over the oceans and must be away from the equator to start spinning because the spin of the earth varies directly with the sine of the latitude. The whole system usually travels at speeds of 10 to 15 miles per hour with sustained wind speeds of 75 to more than 100 miles per hour.

Meteorologists and geographers view them as natural mechanisms that distribute heat from equatorial to temperate and polar regions. The result is a more even earth climate. Also, it has been estimated that as much as one fourth of the precipitation that waters crops and forests and replenishes rivers and groundwater in our southeastern states comes from hurricanes and tropical storms. They are one of nature's most catastrophic phenomena. But they also reshuffle nature, change the coastal environment, and cause plant, animal, and human life to renew itself, creating a new equilibrium, which forever makes the seashore an interesting place.

The Florida coast offers many habitats to sea and land creatures that entice the confirmed beach wanderer. From the sandy shores of the Atlantic coast to the coral reefs of the Keys, from the mangrove swamps of the Ten Thousand Islands to the dunes of the Panhandle, geologic structure and different climates produce homes for innumerable living things. Manila palms, succulent cacti and saltgrass, green turtles, marine worms, exotic fish, and carnivorous corals are only a few of the local denizens.

Between the Keys and the mainland, shallow waters extend from Biscayne Bay through Florida Bay. Inland from the southern tip of Florida to Cape Romano is one of the largest coastal swamps in the world. Birds are still abundant here, especially in the rookeries of the tree jungles, though not so astoundingly numerous as they were before people began to invade their domain. Now that the bulldozers have arrived, the coastal wilderness of Florida is fast disappearing.

In the past, Florida has had a moveable coastline. The state itself is flat and the contour intervals on the maps are far apart. You can travel several miles and find that the elevation has increased only 6 inches. Some 100,000 years ago, when the sea was 25 feet above its present level, the state was covered with water. The sea retreated during the Wisconsin period of the Ice Age. This ancient submersion has given Florida much of its variation of rocks and sand and coral along its shores.

The limestone rock deposits on shallow bottoms were formed by animal shells and calcareous worm tubes. The sedimentary islands along the east coast are the result of sandbars of sediments carried by ocean currents from the north. These islands were once covered with shrubs and trees before man began clearing and dredging. The Keys are remnants of an ancient line of coral reefs that developed on he shallow continental shelf on which Florida rests.

Life on the beaches is so variable that one has to choose which kinds to observe. The birds, of course, are always here. The black skimmers appear in huge airborne flocks. The pelicans drift on the waves. And the shorebirds search for animals that live in mud and sand. If there has been a storm, the beach is strewn with debris; the beachcomber may discover a vast quantity of shells and other sea creatures that have been cast up by the waves.

The collectibles you encounter after a storm depend upon where you are. They may be seaweeds, turtle grass, or mangrove seedlings. They may be horseshoe crab shells, men-of-war, conches, moon shells, starfish, or the empty tubes of worm shells. The list goes on and on, including whelks, wedge shells, pen shells, cockles, turkey wings, and lettered olives. If the tide and day are right, you can collect as many as fifty species of seashells in an hour.

On the west coast, many of the shell specimens are likely to be quite perfect. Sanibel Island, with its long wide beach, is a shell collector's paradise, as we have said. If the tide is right, you may also find some fine shells occupied by hermit crabs at Alligator Point near Tallahassee. Shells on the east coast of Florida are likely to be broken by tides and beach-cruising vehicles. But you can find fine specimens if you persist in your search.

The mangrove islands and their wildlife are a naturalist's world that differs from all other seashore zones. These impenetrable jungles house hundreds of sea creatures. And in the crowns of the trees are the rookeries of thousands of tropical and semitropical birds. The mosquitoes here can drive you crazy, but remembering that they are basic to the food chain that includes tiny fish, shrimps, crabs, lizards, rattlesnakes, fiddler crabs, and other fish can help make them more bearable.

North of the resort strip at St. Petersburg, and extending to the eastern Panhandle, are swamps and muddy bottoms that have defied even the most powerful bulldozers. The shallow silty flats are anathema to tourists. But in these waters are grasses that protect and support stupendous numbers of marine animals, including the young of the game fish that attract tourists and sport fishermen to the open waters of the Gulf.

The corals are among the most beautiful animals of the sea. Their growth patterns vary to form fantastic configurations: tree shapes, masses that resemble large boulders, and multicolored groups that seem like flower gardens, through which gaudy-hued tropical fish wander at random. Scientists study corals because of their physiology, their internal structure, and their relationships to other animals that do not resemble corals at all. Corals are rock builders that have produced islands, reefs, and atolls in tropical seas. The Great Barrier Reef of Australia, the most extensive expanse of lime rock in the world, was built by corals.

The corals of south Florida are the ones most of us associate with the name. You can wade among them in very shallow water, but their sharp edges can cut your feet, legs, and hands if you do not wear the proper apparel. My first introduction to them was by way of a glass-bottomed boat, south of Miami. More adventuresome people get a better view with shallow-water diving equipment or sophisticated scuba diving gear.

The coral reefs of Florida form banks that rest on the shallow continental shelf surrounding the peninsula. Some of the reefs produced in Pleistocene times now form the Florida Keys and the Dry Tortugas. Other reefs are present off the north coast of Cuba, near Veracruz, Mexico, and in the Gulf area of British Honduras. A small stony coral grows as far north as Cape Cod, but it does not form barriers or reefs. Even the rock pools of California have their corals, but these are simple animals, with single cups, which live as solitary creatures instead of groups in great stony masses.

Corals are animals. The stony corals have skeletons of lime with fleshy living parts. Some have stinging cells, which can cause injury similar to that of mild stings from jellyfish. Some kinds of corals have skeletons of horny material, and still others have no skeleton but are fleshy. The many varieties include the star coral (*Montastrea cavernosa*), moosehorn coral (*Acropora palmata*), finger coral (*Porites divaricata*), and sea fans (*Gorgonia flabellum*).

The living coral is a polyp, which in its most simple form is a fleshy tube, open at the top with tentacles. Its structure resembles that of a sea anemone. The outside portion of the polyp secretes calcium carbonate, and as it continues to develop, these deposits form a cup. By branching and further growth, the cup becomes a tube and combines with other polyps to form a chambered rocky mass of great complexity. The numerous polyps maintain a sort of communication with one another. Many corals have greater complexity of structure than can be described here.

Coral polyps are of different sexes. The spermatozoa of male polyps are drawn into the mouths of female polyps, which produce eggs in a skinlike fold, or mesentery. When the eggs are fertilized they emerge and swim for a time by means of cilia. When they become attached to a rock or another coral they gradually develop the polyp form and begin to secrete a stony skeleton.

Sea fans are horny corals that have no stony skeletons. These corals are flexible and present a broad surface to the water currents, which bring minute food organisms to their polyps. Some of these fans are sold in curio stores, but first they must be dried until their almost unbearable odor disappears.

An amazing variety of plants and animals lives within and among the corals. Boring and tube-building worms penetrate the corals, and some build their own limy skeletons. Sponges of many colors find corals a favorable place for attachment. The habitat also supports sea urchins, sea hares, starfish, shrimps, snails, and many smaller animals ranging down to those that are microscopic in size.

The coral environment is a desirable place for fish, offering shelter, protection from predators, and food produced by plants and animals. Angelfish, red parrot fish, and butterfly fish are especially colorful inhabitants of the coral environment. Some fish live by scraping algae from the corals. Moray eels hide in holes and caves in the corals from which they grab passing prey. Other predators include gray snappers and barracudas. Many sedentary animals also adapt themselves to life among the corals by attachment or by forming cups and shallow burrows in the limy structures. Most numerous are mussels, oysters, scallop shells, wing shells, and the reef oyster.

Although most of us never have the opportunity to explore the mysteries of coral formations, we can at least marvel at how they built their vast reefs and structures through the ages. They are vital areas in the life of the sea, building islands where none existed before. And they serve as natural breakwaters that protect coasts from the devastation of violent storms and hurricanes.

When the sea turns from blue to red it portends a time of distress for sea creatures and humans, especially along the Florida Gulf coast. The red tide is a natural phenomenon caused by very small marine algae called dinoflagelates, which periodically accumulate in vast numbers. The developing bloom of algae, which may occur over hundreds of square miles, is transported inshore by wind and water currents, killing fish by the millions. The dead fish are then swept by the tides into malodorous windrows on the gleaming beach sands.

The bottom-dwelling fish are the first to succumb, but nearly all fish are susceptible to the algae's nerve toxin, which inhibits respiration, or a blood toxin, which inhibits oxygen transfer within the fish's body. Humans are driven from the beach by the stinking dead fish and by an aerosol toxin, also produced by the dinoflagellates, that causes choking and eye-burning sensations.

Shellfish that filter the tiny organisms are not harmed themselves, but the toxin is concentrated in their flesh and can cause paralysis or death in people who eat them. These harmful colored tides occur throughout the world and have been recorded since biblical times. It is possible that the seventh chapter of Exodus, which notes that the Nile turned to blood, the fish died, the river stank, and the Egyptians could not drink its water described the effects of a red tide. In other parts of the world, the waters turn yellow, green, or even black with clouds of these tiny unique plant-animal organisms that decimate marine life.

Different species affect different areas of the ocean. *Gymnodinium breve,* for example, ravishes the Gulf coasts. The red tides it causes have been known in Florida since 1844, but the organism was not isolated and described as a dinoflagellate until the 1946–1947 fish kill. Another species, *Gonyaulax tamarensis,* dealt a blow to the New England shellfish economy in 1972. Some thirty people were poisoned by eating contaminated shellfish, as were many birds that fed on them.

The causes of red tides are unknown; however, temperature changes, seawater salinity, behavior of the Gulf current, direction of the wind, enrichment of the sea by nutrients discharged in waters from coastal marshes, and industrial pollutants are all suspected factors. It is known that *Gymnodinium breve* red tides start offshore in areas where a planktonic population is present throughout the year. Although this seed population averages less than 1,000 cells per quart when a "bloom" occurs, they reproduce so rapidly that a quart of water may contain nearly 100 million.

The red tide organism does have a predator—a microscopic luminescent dinoflagellate, *Noctiluca* sp. But the predator does not occur in sufficient numbers to control the bloom. Attempts have been made to control red tide by spraying copper sulfate from airplanes, but the cost is prohibitive because of the thousands of square miles that may be infested. With out present knowledge, red tide cannot be eliminated.

Many investigators agree that the red tide is a part of the balance of nature, although its role in the marine ecosystem is not fully understood. Fishermen do report that shrimp and blue crab catches increase after red tides have caused massive fish kills. It is possible that this increase results from the diminished ranks of the predatory fish. Much research, however, must be done before the public health aspects, the biology, and the ecology of red tides are thoroughly understood.

Salt marshes and tidal channels furnish homes for a multitude of animals ranging from worms, crabs, fish, and snails to birds. This is on the coast of Georgia.

When we explore the waters of the bayou country we find a tremendous variety of habitats where plants, animals, and man are linked together in a struggle for existence. Indians originally lived along these shores. When the white man arrived, the abundant supply of edible wild creatures and the splendid harbors led him to establish cities, such as Pensacola, Mobile, Ocean Springs, and Biloxi. Smaller cities and towns now have many winter and recreational homes for the thousands of travelers who find beach living, fishing, and vacationing attractive here.

The variable physiography and biotic abundance of the shoreline and the inland country along the Gulf offer a wide range of recreational and economic opportunities. Not all the Gulf shore, however, is easily accessible.

The Gulf shore of Alabama is only about 50 miles across. Mobile Bay has 433 miles of bay and open-water shoreline and more than a third of a million acres of estuary open to boat travel and shipping. Dauphin Island, an offshore barrier island, has beautiful dunes and interesting vegetation. Its pure white sand beaches attract large numbers of seashore visitors.

Much of southern Louisiana is a combined gift of the Gulf of Mexico and the Mississippi River, which contributes silt from thirty-one states and three Canadian provinces. This silt has produced the delta and its fabulous but mysterious bayous. Gulf currents have shifted the sediments along the coast. Lowering of the water level has preserved older shorelines as chênières, now tree-covered, elongated ridges of land in the marshes. In the marshes, the tide moves in channels among the grasses and sedges where waterbirds, muskrats, nutrias, and marine creatures spend a part of their lives.

The "bird-foot" pattern of the Mississippi delta channels is the result of centuries of change. In the past, sediments accumulated in distributary channels that emptied into the Gulf. The rising level of silt eventually lifted the level of the river bottom, causing the water to spill over natural levees in various directions. During floods, this breaching of levees gave the water a shorter and steeper outflow, producing new channels and new lobes which, in turn, added to the branching or digitate form of the delta.

In its history, some channels of the Mississippi River have become obscure or have been abandoned. The Atchafalaya River, through which the Mississippi once flowed, has a steep gradient. During floods, it now carries one fourth of the Mississippi's water. Without structures to control it, the Atchafalaya basin, whose channel is only half as long as the present New Orleans–Plaquemines channel, could once more carry the water of the Mississippi.

One has to detour off the interstate highways to see and enjoy the charm of the Delta with its magnificent cypress-tupelo swamps and its mysterious bayous. Some of these are difficult to see on foot because the ground is soggy, and there is an almost impenetrable vegetation. In addition, cottonmouths and rattlesnakes are an ever-present threat. Other bayous can be explored only by boat, preferably with someone familiar with the intricate puzzle of water channels in the swamps and marshes. The native Acadians, the French-speaking descendents of settlers who were driven from Nova Scotia by the English more than two centuries ago, know these waters best.

The bayous are sleeping waters that slowly move hither and yon, maintaining a wet misty balance between land and sea on the sinking fringe of the continent. The waters of the bays, marshes, and swamps in the bayou country go in different directions, depending on flow from rivers and pressure from the tides. The channels are deep or shallow, wide or narrow, and so intricately connected that only a bayou man can travel them with safety.

Many birds find refuge and food in the bayou waters. Egrets and other waders live here. Also present are the American anhinga, magnificent frigatebird (nonbreeding), black-necked stilt, Forster's tern, least tern, and black skimmer. The fulvous whistling duck resides near Crowley, west of Abbeville, as well as in southern Texas.

Louisiana has 5 million acres of marsh where raccoons, minks, and otters abound. Most numerous, though, are the muskrats. Trappers bring in nearly ten million muskrat pelts each year. Nutrias, South American fur-bearing rodents that escaped from pens during a hurricane and established homes in the marshes, are also included in the catch.

Muskrats have always intrigued me because they are one of the most adaptable mammals of creeks, rivers, and marshes. When I was a boy, I trapped them in the cattail marshes that bordered the Missouri River. At one time, the pelts were worth $2.50 each. The twenty-two muskrats I caught and skinned one night in a pond on our Nebraska farm made me the envy of my classmates and the richest sixth grader in the school. Years later, as a biologist, I no longer trapped them, but their adaptations to differing water environments continued to fascinate me.

When they live in creeks and rivers they build tunnels, which they enter below the surface of the water. In marshes and shallow lakes, their houses look like stacks of hay, although they are made of grasses, sedges, cattails, and reeds. All during the autumn, the V-shaped wakes of muskrats can be seen as they swim with nest material trailing beside their bodies. The mud hummocks on which the animals place vegetable debris are covered deeper and deeper until they reach a height of several feet. These houses protect the muskrats from marauding foxes, hawks, and owls.

Muskrats insulate the rooms inside their houses with grasses and weeds. In northern climates, the mound of mud and vegetable matter around the inside room freezes solid when the lake becomes covered with ice. The muskrats do not hibernate in winter. They can swim under ice for considerable distances, especially if there are air holes or large air bubbles through which they can breathe. I once chased a muskrat for nearly thirty minutes as it swam in mud channels beneath three inches of crystal-clear ice on a Wisconsin lake. The animal could see me and I could see it but it never came to an air hole. Somehow, it managed to find air for breathing.

A great concentration of muskrats occurs in the bayous of Louisiana and the marshes along the Gulf coast. In these southern marshes, the muskrats have many enemies other than man. The young, the first of which are born in spring, fall prey to snakes, marsh hawks, owls, and minks. These latter animals are fierce predators and will invade muskrat houses to kill both young and adults. At the end of the food chain are the alligators, which lie in wait for any hapless creature that comes within reach of their jaws. But, like, mice, muskrats have a high rate of reproduction. They are always plentiful in the vast southern marshes.

The story of the Texas Coastal Bend is not an entirely happy one. The Gulf, bay, and estuary shores total 2,498 miles at mean high-tide level. The federal government owns 16 percent of the shoreline, and 2 percent is owned by the state, by counties, and by cities. The remaining 82 percent is privately owned, but people and industry in this portion of the shore are not all inclined to permit public access to the beaches. The Open Beach Law of Texas is supposed to protect the public's beach rights. Even though the statute has been amended, it is still controversial and does not work in practice.

The physical characteristics of the Texas shore should allow almost unlimited opportunity for recreation and other forms of seashore enjoyment. Beaches of fine sand, shells, and shell fragments slope gently upward to sand dunes that reach 40 feet in height. Coastal bay marshes are numerous. And waves formed by the prevailing south to southeasterly winds have produced bluffs up to 40 feet high on the north and west shores of bays.

Hurricanes have littered some of the shore with rotting hulls of shrimp boats and rusting automobiles. The Rio Grande is only a trickle where it enters the sea, since most of its water has been damned upstream for cotton field irrigation. A Conservation Foundation Study, *The Decline of Galveston Bay,* states, "Houston, the Ship Canal, and Galveston Bay [display] a dichotomy between the national concern for improving and protecting the quality of the environment and a reliance upon an economic system that treats pollution as an external cost of doing business and that has prospered and grown at the expense of environmental quality."

Author examining muskrat house in a marsh pond. These small animals expend a great deal of energy in building their winter homes.

Houston is Boom Town, U.S.A. There is the story of two Texans discussing the size and worth of their holdings. The rancher says, "My spread is so big it takes all day to fly from one end to the other in a helicopter just to count the cows." The other Texan says, "My spread is fifty-seven acres." When asked where his spread is, he replies, "Downtown Houston."

As late as 1836, Houston was an uninhabited swamp. However, things changed dramatically, for in 1902 oil was discovered. In 1914, the U.S. Army Corps of Engineers dredged a ship canal from the Gulf of Mexico to Houston, which is now the third largest port in the United States. The canal conveys not only ships but wastes, chemicals, minerals, and organic material. This raw, untreated, unchlorinated sewage is discharged into Galveston Bay. Pollution has closed large areas of the bay to harvest of shellfish. Fish and other marine life have been killed. And thermal pollution from new electric generating plants is destined to raise the water temperature, possibly to limits that will be lethal to additional marine life.

But the end is not yet in sight. A grandiose plan to dig a trans-Texas canal to intercept all tributary drainage to all coastal marshes is now afloat. By the year 2010, the rivers of Texas would no longer flow to the sea. Instead, they would be diverted to eighteen reservoirs to supply the canal with water for projected municipal, industrial, and irrigation requirements. The demand would reduce by one half the average water flow of 26½ million acre feet now reaching the Texas estuaries. The Texas Basin project would account for one third of this reduction.

The two Padre Islands are still attractive recreation areas. First mapped by Alonso Alvarez de Piñeda, they were known as the White Islands. In 1553, part of a twenty-ship treasure fleet was wrecked here by a hurricane. Some 300 men, women, and children survived the wreck, only to be killed and eaten by the Karankawa Indians. Modern treasure hunters still come to the Padre Islands to search for gold. Occasionally, a doubloon or a piece of ancient chain mail is found.

South Padre Island has some truly wild stretches of beach. Its National Seashore, 69 miles of relatively undisturbed beach and dune land, is a great seashell-collecting area. And one never knows what flotsam or jetsam will be brought to shore by the next tide or the next storm from the sea. But, aside from the National Seashore, the fragile ecology of North Padre Island is being endangered by human exploitation. Bulldozers remove the dunes and permit the sand to blow away. Condominiums, fast-food joints, adults, children, cats, dogs, and BB guns all take their toll on the environment. Wildlife here still includes some 300 species of birds that inhabit the dunes or pause for a time during migration. Especially abundant are gulls, terns, brown pelicans, snowy egrets, and great blue herons. On the salty lagoon, Laguna Madre, between Padre Island and the Texas mainland, redhead ducks form great rafts. In the island marshes are geese and roseate spoonbills. The plants of the dunes include sea oats, wild morning glories, cacti, and various shrubby species.

Still unexploited are Mustang, Saint Joseph, and Matagorda islands. The last-named shelters the vast Aransas National Wildlife Refuge, north of Corpus Christi, where the whooping cranes winter. In this vast marshy area the late Dr. Clarence Cottam counted 450 bird species. The last time I was in this part of Texas I spotted one of my favorite shorebirds, the long-billed curlew. The inland areas of the state were covered with snow and ice, but, despite the cold weather, the birds were busily searching for crustaceans in an open marshy area. They brought back memories of long-billed curlews I had known during summers on the prairies of Colorado and around the desert lakes of Oregon.

These largest of shorebirds used to nest on the Central Plains Experimental Range north of Greeley, Colorado. They always showed much distress when we approached their nests to look at their greenish-olive eggs. Grasshoppers were generally abundant and comprised a fair share of their diet. In the salt marshes of Florida, Louisiana, and Texas, their winter diet includes mollusks and seeds. Whenever I see them on the prairie or along the Gulf coast, I recall the great slaughter, when market hunters shot into the enormous flocks and killed them by the thousands. Now they are few in number and are protected by law.

There are still many other biologic and recreational treasures along the Texas Coastal Bend. These resources of wildlife, sandy beaches, and natural beauty are there for the American people, who should understand that even in their natural state these coastal areas make important economic contributions. The waters supply about 200 million pounds of fish and shellfish annually. And they support more than 7 million man-days of sport fishing. They can continue to produce these products and pleasures only if men have the intelligence and the desire to preserve them as they are.

Totem poles are vanishing symbols of the Indian culture that formerly existed along the Northwest coast.

10

Man and the Seashore

INDIANS WERE LIVING AT THE OCEAN'S EDGE when white men first landed on Atlantic, Gulf, and Pacific shores. Now we no longer see the original inhabitants, such as the Nootkas, Chinooks, Tchefunctes, and Roanokes, some of whom established their cultures at least 10,000 years ago. There were many tribes, some with small numbers of individuals, and the different groups had many ways of adapting the gifts of the land and the nearby sea for their use. Only a few tribes now remain, most of them located along the Northwest shore.

Most of what remains of Indian culture consists of shell middens along some of the Atlantic and Gulf coasts, although totem poles and artifacts still appear in the Northwest. Legends of former beliefs still remain, and a few customs, like the potlatch, still persist. The potlatch was a gigantic feast that was held when a great leader died and his title was inherited by another Indian of high social stature. At the time of the potlatch, a totem pole was carved in honor of the host, with animal images and crests depicting his lineage and status. Gifts were exchanged near the end of the ceremonies and the great feast was finished. The potlatch was celebrated from the Strait of Juan de Fuca northward to Alaska.

The seashore Indians lived with, not against, nature. The bounty of the sea and the flora and fauna of the land provided them with food, tools, and shelter. Their language and culture varied with their surroundings. Some tribes used shell money. Mollusks were commonly used for food. Seagoing Indians engaged in whaling expeditions, fished for salmon, caught halibut, and killed sea otters in the kelp beds. Indians on Gulf shores wore little, if any, clothing. On northern shores of both sides of the continent, they built strong houses of wood and warmed themselves in winter with fire.

The river-dwelling Indians, such as the Chinooks of the Columbia River valley, traded with inland Indians, some of whom brought hides from as far away as the Great Basin and even the Great Plains. Other items of trade included fish, hides, shells, canoes, and even slaves. The Chinook tribe was estimated at 5,000 individuals when Lewis and Clark met and lived with them at the mouth of the Columbia. As other whites arrived, the Indian language merged with French and English. And it was the white man's alcohol and diseases that finally did them in.

History is not clear on the order in which Indians arrived on the American continent. Nor is it clear how the different tribes developed and why they settled in specific territories. The relative abundance of food and building materials in the environment undoubtedly shaped their habits and some of their cultures. The inhabitants of the Northwest coast, particularly those of British Columbia and Alaska, were seagoing Indians. The Nootkas and Kwakiutls had an abundance of fish and other sea creatures for food and consequently had leisure time in which to feast and build their houses of cedar.

The Yuroks and Hupas of northern California were hoarders, possibly because of a lesser abundance of food. But their diet did include acorns and other vegetable materials, as well as fish. Since they were primarily river dwellers rather than seashore inhabitants, they ate salmon and steelhead, fish that spawn in rivers and then migrate to the sea.

Many tribes lived along the Pacific coast when the white men first arrived. Notable among these were the Nootka of Vancouver Island, the Makah of the Olympic Peninsula, the Tillamook and Siuslaw of the Oregon coast, and the Indians of the California culture. Some anthropologists believe this culture represents one of the oldest aboriginal groups in the United States.

When Ceremeno arrived in Drakes Bay in 1595, the Coast Miwok of the California culture were among the first western Indians to be seen. These Indians resided primarily in what is now Marin County, as well as in parts of Sonoma and Napa counties. They built villages at the edge of the sea and used creatures from sea and land, including fish, birds, and various kinds of mammals, including deer, rabbits, and dolphins, and many varieties of plants, to supply their needs. The shell middens left by these Indians also indicate their extensive use of clams, abalones, chitons, and barnacles.

The construction of Mission Dolores in San Francisco, beginning in 1776, marked the beginning of the decline of the Coast Miwok tribe. As with so many Indian tribes, the missionary's ignorance of their culture, compounded by epidemics of measles and syphilis, made them virtually extinct in the San Francisco region by 1816.

The historical heritage of the Tillamook, the southernmost of the Northwest culture group, has been better preserved than the history of the other tribes of the Pacific coast. Meager descriptions of this tribe were written by Captain John Meares and Robert Gray. But the journals of the 1805–1806 Lewis and Clark Expedition recorded in detail their contacts with the Tillamook, inhabitants of this region for possibly 10,000 years.

The Indians of the Gulf coast built their campsites and left their shell middens and cultural remains in what are now near-sea-level lakes, marshes, swamps, and bayous. Their territory was a movable landscape since the rivers frequently changed course and the shore-line advanced or retreated under the influence of the restless sea.

Potsherds, shells, and animal bones are now found on ridges or chênières in Louisiana where the Tchefuncte Indians lived some 2,000 years ago. Potsherds and other artifacts have enabled archaeologists to identify other early occupants as well, including the people of the Troyville Age, from about A.D. 600 to 900, and the Coles Creek Indians, from about A.D. 900 to 1,200. The Louisiana Indians of the Sabine River country were the Atakapans. This now-extinct tribe, subsisting on roots, nuts, fish, and small animals, led a scavenger type of existence. An even more primitive tribe, the Karankawa, lived primarily from Galveston Island to Padre Island. These Indians practiced ritualistic cannibalism and lived without clothing or houses. They were practically exterminated by the end of the nineteenth century, and the few remaining survivors moved to Mexico.

On the Atlantic coast, tangible evidence of Indian habitation dates back about 4,000 years, although some regions may have been inhabited as long as 10,000 years ago. Shellfish were an important source of food, as indicated by the presence of shell middens; corn, beans, and pumpkins were also cultivated as staples. The Roanoke

Indians on the Outer Banks of North Carolina wore clothes probably made from the skins of deer. Venison, fish, and fruits were cooked or eaten raw. Their houses had several rooms.

In most instances, the Indians who lived along the seashore had an abundance of natural products. While they did leave their imprint on northern lands by setting fires that destroyed forests and drove game out into the open, they also built villages and settlements that were surrounded by cornfields and melon patches. They did not graze the native forage to near extinction with cattle, horses, sheep, and pigs, nor did they pollute the estuaries, drain the marshes, or change the conformation and beauty of the shoreline. They left this kind of desecration to the white man.

The history of settlement, beginning with explorations of all our coasts, is a long one. Some of us may remember a few of the details from our history lessons in school; there are many books that can help fill in the gaps if you are interested in outside reading. The main point here is that early arrivals of the Spanish, French, English, and Russians on our shores spelled the beginning of the end for Indian settlements that had been here for centuries prior to the landing of Columbus.

The explorers had little effect on our seashore itself. These earliest white men came to search for gold and silver, to discover northwest passages to the Orient, to found missionary colonies, and to take possession of new lands for the greater glory of their Majesties, the kings and queens of Europe. But they laid the foundation for the eventual expansion of settlement of our nation. Half of all the people in America now live within 50 miles or less of the seashore.

Some of the early settlements were failures. Soon after Columbus, white men arrived at Zacatula, northwest of what is now Acapulco on the Pacific coast. Cortez attempted to found a colony at La Paz in 1535, but the project was unsuccessful and Baja California was left undisturbed until nearly 150 years later. Eusebio Francisco Kino, of the Society of Jesus, tried to establish a settlement at La Paz in 1683. The Indians revolted and Kino fled to the mainland. In 1767, the Jesuits were expelled from all Spanish possessions; the Franciscans moved northward to California; the Dominican Order of Preaching Friars, which looked unfavorably on nudity and polygamy, also failed when wars erupted and the white man's diseases decimated the Indians. Tourism along the seashore, as we now know it, did not start until about 1920 in Baja California.

Bodega Bay, California, was discovered by the Spanish explorer Juan Francisco de la Bodega y Quadra on October 3, 1775. But the Russians became the principal settlers there, raising livestock and vegetable crops. Before they moved to Sitka, Alaska, in 1841, the Russians nearly exterminated the sea otter and seal population. Thus, man began his tinkering with the wildlife of the seashore.

The search for gold and empire continued for 200 years or more. In 1542, Bartoleme Ferrelo explored as far north as Oregon. Sir Francis Drake reached the Rogue River in 1579, turned back because of foul weather, and made a landfall near San Francisco. Captain James Cook, on his voyage of 1776–1780, missed the Columbia River and the Strait of Juan de Fuca. The Greek explorer of that name, in the service of Spain, claimed it was rich with gold, silver, and pearls. He little dreamed that the magnificent forests of spruce and fir on the mountains inland from the sea were the real treasure that would bring civilization and industry to the Pacific Northwest.

Some of these explorers of the Pacific coast met the Indians and bargained with them for furs and traded metal objects for other products of the forests and seashore. They took notes, wrote accounts of the native inhabitants, but gave little thought to the fabulous fish resources in the rivers that flowed into the Pacific Ocean.

Nauset Light, Cape Cod. Lighthouses overlook the "graveyards of the sea," but ships still sink in spite of these warning beacons.

Even Captain George Vancouver, in 1792, ignored the Columbia River, deeming it unworthy of exploration. When the conquest for empire finally began, it came from the interior of the continent instead of from the seashore.

The Gulf coast states, likewise, were originally settled from inland. In 1846, after General Zachary Taylor defeated the Mexican army and Texas was admitted to the Union, settlers moved in with their crops and livestock. The unlimited grasses of the prairie were the main treasure for the ranchers with their great herds of cattle. Between 1840 and 1860, the only coastal industry of importance was the capture and sale of green turtles. These creatures no longer inhabit the Texas bays or come to the sandy beaches to deposit their eggs.

Eastward, the Spanish established St. Augustine in 1565. Their fleets also explored the Caribbean, Mexico, and Central America all during the sixteenth century. The French, too, were busy exploring the northern and southern coasts. Not until 1607 did the English establish their first permanent settlement at Jamestown. Earlier, in 1548, Captain Arthur Barlowe, an English explorer, had visited the Outer Banks and described the Indians living there as friendly, with a relatively sophisticated culture that included five-room houses. The country was productive, with an abundance of grapes, deer, hares, wildfowl, and fish.

The Croatan Indians of the Outer Banks of North Carolina intermarried with the English settlers and cooperated in farming, fishing, and hunting. Soon, horses, cattle, sheep, and hogs were introduced and timber cutting began to have its effect on the stability of the sandy lands. Hurricanes battered Hatteras Island, as they still do today. But it was the timber cutting and overgrazing that loosened the sandhills and allowed the winds to change the design of the land. That was only the beginning of man's devastation of the virgin seashore in America.

However, man does not always win in his attempted conquest of the sea and its shores. Sometimes the sea retaliates. When storms are raging and tides are high, they cast the wrecks of ships upon the rocks and beaches of the world. The sea also removes roads and habitations that have been built too near the fragile edge of continents. It chews away rocky cliffs, causes islands to disappear, and produces landslides on eroding shores. Given enough time, the sea may yet prevail, as it has in the past, by invading the continents themselves.

Recently, there has been much discussion about world climatic changes that may be brought about by man's increasing use of fossil fuels. Scientists are not saying that this will occur during our lifetimes. But the "greenhouse effect" caused by overloading the atmosphere with carbon dioxide from burning oil, coal, and natural gas may allow heat from the sun to reach the ground and prevent its radiation back into space. The ultimate effect could be that the ice sheets at the poles would melt, causing sea levels to rise dramatically —inundating all our shores and many of our habitations. At at recent annual meeting of the American Association for the Advancement of Science, John Mercer of Ohio State University said that we are presently in a worldwide warming trend that may last for 1,000 years, but the rate of warming could be intensified by our use of fuels.

The sea has won many smaller battles with man, as well. The evidence appears from time to time along the seashore. Point Conception of the California coast has been called the Cape Horn of the Pacific. Padre Island in the Gulf has had its share of wrecks, beginning with the pirate ships of earlier times. And the waters near Cape Hatteras have long been known as the graveyard of the Atlantic. Beaufort Inlet to the Virginia coast has long been strewn with remnants of lost vessels.

A violent hurricane from Jamaica struck the North Carolina coast on October 9, 1837. Three ships were sunk and the shore was strewn with bodies. Henry David Thoreau, who visited Cape Cod in 1849, 1850, and 1855, described how the bodies of emigrants from Galway, Ireland, were collected in large boxes after a great storm had wrecked their ship. Two or more children or a parent and a child were placed in the same box. But the men of Cape Cod had become so inured to the tragedies of shipwrecks that at the same time some were picking up seaweed for fertilizer and others were bound for the beach with game bags and hunting dogs.

The story of the sea's victory over men and their ships is almost endless. The sea lanes past the Outer Banks have seen ships of every kind. The commerce of the world has passed there for hundreds of years in schooners, full-rigged ships, sloops, naval vessels, and cargo ships. When the storms come to that dangerous coast they bring the battle of life and death. And the Cape Hatteras Lighthouse does not save them.

The first lighthouse on Cape Hatteras was authorized by the U.S. Congress on May 13, 1794. It was completed about 1802 and, in 1854, extended in height to 150 feet. Then, a new one, 180 feet tall, was built. It was struck by lightning in 1879. The ocean was half a mile east in 1854, but the water's edge was almost at the base of the lighthouse by the 1930s. It has been repaired and reactivated and its beacon is visible from 20 miles away in clear weather. The treachery of the sea in its front yard derives from the ever-changing shoals and channels which extend out from Cape Hatteras and the turbulence of warm waters of the Gulf Stream meeting cold Arctic waters from the north. Dense and impenetrable fogs add to the hazards of navigating here.

The story of shipwrecks around Point Conception, that sharp corner on the California coast, is the same. Winds from the northwest swirl around the point and down the Santa Barbara Channel. When a Santa Ana blows from the east, it brings new hazards to mariners. The shipwrecks have included vessels loaded with people and gold, passenger streamers, and U.S. Navy destroyers. More recent wrecks were the *Harvard,* a luxury ship, off Point Conception in 1938 and the *Chickasaw,* at Santa Rosa Island in 1962.

The sea has one additional force of destruction which neither man nor his structures can withstand—the tsunami. This is a seismic sea wave that is set up not by the tides but by submarine earthquakes and landslides. Tsunamis tend to originate in deep ocean trenches. Some are caused by vertical displacement of huge blocks of the earth's crust.

Since these giant sea waves may be only a few feet in height and 100 miles or more in length, they are often barely perceptible to ships on the ocean. In the open ocean they may travel at speeds of 500 knots. Their destructiveness comes when they flow up sloping shores and build up waves of disastrous proportions.

In 1946, a seismic wave originating in the Aleutian Trench caused damage totaling $25 million in Hawaii—more than 2,000 miles away. Another recent wave from Alaskan waters caused damage to California shores. One of the great dangers of seismic waves is that their arrival may be preceded by the recession of water from the shore. This may lead people to believe, falsely, that an extremely low tide has offered an opportunity to explore the edge of the sea. When the crest of the wave arrives, people and small boats are engulfed and destroyed.

Man's onslaught against the shore is often highly visible: the construction of houses, hotels, resorts, and marinas, the draining of marshes, the leveling of dunes, and the erection of coastal fortifications, including rock groins, concrete seawalls, and rubble revetments, are everywhere. But man's impact against the shore is also invisible and insidious beneath the surface of the sea, such as when littoral currents carry away beach sands, pollutants kill fish and alter the ecology of the shore, or thermal pollution follows the construction of power plants.

The principal offenders against the beauty and usefulness of the seashore are not the municipal and port expansion authorities, the real estate promoters, and the oil spillers; they are the homeowners who demand protection from wave action and the industries that have no motive other than profit. We, the people, are offenders, too,

Beach houses are a familiar sight in many areas from Maine to Florida, and to some extent from California to British Columbia. The owners no doubt enjoy them, but they do not enhance the beauty of the coast and they limit access for millions of people who would like to visit our seashores.

in that we have not generally recognized that the seashore belongs to all of us and that individually and through organized effort we can bring pressure in high places to preserve what remains of one of our greatest natural heritages. It can be done. California has a coastal plan. Massachusetts also has a wetlands plan. Other states are beginning to take action.

What have we done to our seashores? We have ignored the swift tidal currents that move the littoral drift and deposit it here and there. Jetties, groins, and seawalls built to protect harbors, hotel fronts, and recreational beaches only accelerate the process. Soon, the sandy front yard begins to disappear. Then the engineers begin to haul in more sand, a treadmill process that assures them of employment in perpetuity, or until the sand supply runs out.

The two-mile-long shore at Surfside and Sunset Beach south of Los Angeles began to lose its sand in 1944. Some 200,000 cubic yards of sand was hauled in. Two years later, a million cubic yards was required. Five years later, the sand had to be replenished. Sand also has to be pumped out from behind harbor jetties. The story is the same elsewhere, including Miami, Florida, and Atlantic City, Cape May, Ocean City, Asbury Park, Poverty Beach, and Shark River Inlet in New Jersey.

The supply of sand is not endless. Beach erosion control in combination with littoral currents may result in a conveyor belt that carries sand into deeper water. Frenetic dam building by engineers also has cut off the supply of sand from rivers that originally nourished beaches. On Atlantic shores, resort promoters have destroyed the dunes that slowed the fury of the winds and the movement of sand from the sea. When the littoral currents chew away the shoreline, cottages fall into the ocean; the owners are the ones who suffer, since by this time the promoters have usually deeded the ocean strip to the local community.

Many have conspired against the shoreline in many other ways. About 50 percent of the estuaries along the California coast have been dredged and filled—and destroyed. But a more subtle change has resulted from man's inadvertent introduction of biota from other continents to our ocean, estuaries, and seashores. Wooden ships long ago transported fouling and boring organisms to our shores. These exotic emigrants include barnacles, snails, hydroids, tube worms, flatworms, and commensal protozoans.

The methods whereby intertidal invertebrates are introduced

from foreign waters are many. Ship ballast, which may be either water or stones and debris from beaches, discharge species that may come from New England to California or even from Japan. Lumber ships apparently introduced the Chilean beach hopper into San Francisco Bay. The Oriental shrimp also appeared there in the 1950s.

Some of the introduced animals are destructive. *Light's Manual,* edited by Ralph I. Smith and James T. Carlton, lists some of the invertebrates that have been transported with adult and seed oysters to California from Atlantic waters and from Japan. The oyster drill, a snail that subsists on oysters, is especially destructive where the commercial oyster industry exists. Other introduced species include sponges, coelenterates, polychaetes, mollusks, and crustaceans. Commercial bait and seafoods are commonly packed in algae for importation. When the algae are discarded into bays they carry with them such creatures as the Atlantic periwinkle. Even marine laboratory scientists allow exotic species to escape into seawater with little regard for the ecological consequences.

Among the serious offenders of seawaters are industries that change the chemical quality of the water and increase its temperature. Pulp mills on the Northwest Coast formerly produced many harmful wastes but now increasingly comply with protective regulations. Desalination plants are common in tropical and desert areas where fresh water is limited in supply; however, the heated brine and toxic effluents can be especially destructive to biotic constituents. The plant at Key West, Florida, however, receives its water from wells and discharges its toxic effluent into an artificial basin instead of into the sea.

A threat to the environment of the Gulf wetlands is the plan to store imported oil in salt domes in southern Louisiana and Texas. Storage cavities are made by dissolving the salt with fresh water under pressure. Disposal of the resulting brine, which has a salinity five times that of seawater, could be catastrophic to plant and animal life. Since the waters of wetlands move slowly, the contamination can remain for long periods of time. It may also penetrate freshwater aquifers and contaminate fresh water that is now used for drinking and other domestic uses. Geologists suggest that the alternative is to use salt domes far inland in Louisiana and Mississippi.

In pumping vast amounts of crude oil into underground salt caverns, the Department of Energy has found that the cost of the

Strategic Petroleum Reserve keeps increasing beyond previous estimates. Too much pressure from pumping oil into salt domes has
resulted in oil spurting from the caverns, catching fire, and burning
expensive machinery, not to mention people. And, not least of all,
the engineering procedures for withdrawing the oil when needed
have not been satisfactorily solved.

Probably the greatest enemy of the seashore and its living creatures is the black tide that follows an oil spill from a tanker carrying
crude oil. These catastrophes occur with disconcerting regularity. A
devastating oil spill came from the U.S.-owned tanker *Amoco Cadiz*,
along the coastline of Brittany, France. Sixty-eight million gallons of
Arabian crude oil, churned into a slimy muck by the tides, settled in
the silt on the ocean floor or spread from the rockbound tanker for
100 miles, crippling the region's fishing industry, tourist trade, and

Industry is part of man's conquest of the seashore. The air, water, and landscape
are polluted by power-plant complexes, urban growth, and highways bringing
trucks and automobiles to the ocean's edge. All of these eliminate seashore
recreation for millions of people.

seaweed-chemical business. The full extent of the ecological damage is still not known.

The magnitude of the oil spill problem is immense. From 1969 to 1974 there were more than 500 tanker accidents. On August 9, 1974, the supertanker *Metula* spilled oil from a load of 196,000 tons on Tierra del Fuego. The emulsion layer spread over 50 miles of beach. Cormorants, gulls, terns, penguins, albatrosses, and other birds died. The local fishing industry was adversely affected. Sea organisms were smothered, and it is estimated that some effects in estuaries will be visible for twenty years.

The oil spills have continued. The *Argo Merchant* ran aground northeast of Nantucket Island, December 15, 1976, and spilled 7.6 million gallons of oil. On December 17, the Liberian-registered tanker *Sansinena* exploded in Los Angeles harbor; nine people died and fifty were injured. On December 24, the *Oswego Peace* spilled 2,000 gallons of oil near Groton, Connecticut. On December 27, the *Olympic Games* ran aground in the Delaware River near Phila-delphia, spilling 133,500 gallons of oil and fouling the shorelines of three states. On January 5, 1977, the U.S.-registered *Austin* spilled 2,100 gallons of oil in San Francisco Bay while loading at Martinez, California.

The cleanup job after an oil spill is expensive. On June 23, 1976, the barge *Nepco 140,* with 7 million gallons of heavy fuel oil, went aground in the American Narrows near the Thousand Islands Bridge. One-half million gallons of oil moved down the St. Law-rence Seaway, causing damage to beaches, private property, ducks, geese, and other wildlife. The cost of the cleanup was $6.5 million.

Cleaning a beach is a long and arduous process. Much manual labor is needed, and finding a disposal site for the oil and debris presents a major problem. If the spill is near the shore, rapid action is required. Sometimes, the oil may be contained by booms and by barriers made of straw packed in chicken wire. Huge spills, such as the one made by the *Amoco Cadiz,* seem to be impossible to contain.

The lasting impact of oil spills has not yet been adequately stud-ied. The initial evidence of damage tends to be forgotten quickly, since dead birds, fish, and other animals soon disappear from sight. The numbers of dead animals that sink to the bottom are unknown. Migratory fish may be absent at the time of the spill, but are killed later when they return to the scene of the disaster. Since oil retains its toxicity for months and biodegradation may take years, oil particles that sink into bottom sands may result in ecological damage to the food chain that is unrecognized by casual observers or even ignored by competent scientists.

Saving oiled birds presents a large problem and requires noble effort on the part of people willing to save wildlife. Probably the best for coping with this problem is the recent guide, *Saving Oiled Seabirds, A Manual for Cleaning and Rehabilitating Oiled Waterfowl,* distributed by the American Petroleum Institute, Washington, D.C. This manual describes the physical effects of oil on birds, collection and initial treatment, care of birds in captivity, cleaning and drying, preparation for release, common medical problems, food recipes, and supplies and equipment. Cleaning an oiled bird is no job for an amateur. This booklet is especially useful because it is based on evaluation of many techniques tried on thousands of birds.

Cleaning an oiled duck, gull, cormorant, or albatross requires love, patience, good care, and equipment. The deoiling process must be followed by drying, feeding, antibiotics if necessary, a chance for the bird to preen its feathers and swim under supervision, and, finally, tagging and release. Record keeping also is necessary. But the time and effort expended have many compensations.

Our attraction to the coastal zone issues from a multitude of needs and desires. There is the demand for living space in exotic environments and the need for recreation away from the miasma of cities and crowded suburbs. Next is the desirability of the shoreline for commercial and industrial activities including power production, shipping, waste disposal, mining, and oil production. Food production from the harvest of fish, shrimps, oysters, and crabs is dependent not only on the sea but on the shoreline nurseries where many of these creatures begin their lives and later return to spawn. And finally, special government uses require space for military, Coast Guard, and NASA bases, and for parks and national seashores.

The harvest from the shore and the nearby sea is great since ocean water is a vast mineral mine. Every gallon of seawater contains more than a quarter of a pound of salt. Magnesium hydroxide is extracted from the sea. It is precipitated to form milk of magnesia and subjected to electric current to obtain metallic magnesium. Bromine has medical properties, is used in photography, and in antiknock gasoline. But these are only a part of the storehouse of dissolved substances in the sea. Copper, gold, uranium, and more than seventy other elements occur naturally there. But the processes for extracting most of these are presently too expensive to be economically feasible.

Drugs from the sea are now receiving increased attention from chemists, microbiologists, and pharmaceutical researchers. From time immemorial, it has been known that people who ate seaweeds had a low incidence of goiter. This is because of the iodine content of these plants. Through the centuries, seaweeds and sea animal products have been proposed as nostrums for the cure of toothache, dropsy, abscesses, menstrual difficulties, and even cancer. Some of these cures are now becoming realities.

Antibacterial substances have been isolated from phytoplankton. Alginate wool, from brown algae, is used by dentists to stop bleeding. The survival time of mice afflicted with lymphatic leukemia has been extended by extracts from sponges. Hormones from gorgonians, an order of corals, have been found to stimulate smooth muscle and to reduce blood pressure. A potent cardiac stimulant, eptatretin, which acts similarly to digitalis glycosides, is obtained from the Pacific hagfish. Even a tube-dwelling polychaete, a segmented worm, furnishes extracts of thelepin, which acts in a manner similar to that of griseofulvin, which is used to treat infected toenails.

Many of the medical compounds extracted from sea creatures have chemical structures unlike any compounds found in terrestrial plants and animals. The great value of these substances is that they may serve as models for development of new drugs. It has been speculated that many creatures in the sea have been there so long that they have developed immunity to ailments that affect terrestrial animals. The horseshoe crab, for example, produces substances that may be useful in medicine. This animal has certainly been around long enough to develop antidisease substances in its body; its ancestors first crawled out of the sea over half a billion years ago.

The seashore offers many other creatures useful to man. Oysters, crabs, scallops, abalones, and a large variety of fish are good to eat. On the Pacific coast during the winter months, lingcod move from the ocean depths into shallow rocky areas to spawn. They reach weights of 30 to 40 pounds. Other species that make great catches are cabezon, striped perch, and greenling.

A typical Columbia River fish is the smelt, which also goes up the Umpqua River in Oregon and up the Cowlitz and Kalama rivers in Washington. The smelt run is unpredictable; from 1958 to 1970, no smelt swam up the Columbia and into the Sandy River near Troutdale, Oregon. But when the run is on great numbers of people congregate. Some men stand on the riverbank and dip with long-handled nets, while others wade into the water with buckets, garbage sacks, and kitchen strainers. Commercial smelt dippers commonly dip from boats. But I remember seeing a woman in formal dress standing in high-heeled shoes on a log that projected into deep water. After the run is over, a fried smelt-eating contest attracts many contestants. And believe it or not, the ladies are frequent winners.

The sardines, which are similar to smelt in size, no longer come to Cannery Row at Monterey. Greed and overfishing caused their demise. The cannery whistles are silent and the cutting tables are bare. The Chinese shacks near the cannery have long since burned down, as have half the cannery buildings. John Steinbeck would be saddened if he were to see the weathered and rusted remains of the venerable old buildings that still remain. The people were sad when all the pilchards were caught, canned, and eaten. I felt sad the last

Marinas provide anchorage for fishing and recreation boats. They occupy miles of sea and bay shores but they do not create oil spills, such as those from large tankers.

time I looked at the rusted iron and the weathered gray boards and wondered which other harvest of the sea—the whale, the oyster, the blue crab, or the shrimp—would be next to go. All was quiet now. Even the cormorants stood on the rocks in the bay without moving.

Commercial fishing in U.S. waters amounts to more than 4 billion pounds annually. Production of hard clams in the Middle Atlantic states regularly exceeds 6 million pounds. In the Gulf of Mexico, virtually the entire catch is comprised of species dependent on estuaries and marshes. Sport fishing supports many millions of man-days along the seashores of the nation.

For individual sport and enjoyment, the variety of seashore creatures is large. Clam diggers must do the actual digging and capturing. There are bag limits and regulations that differ with the species and from state to state. Female crabs must be returned to the water in Oregon and other states, and there is a size limit on males. For the adventuresome, abalone diving requires skill in underwater swimming, a bar to pry off these seagoing steaks from rocks, and a mallet to pound the slices of meat before they can be broiled over a beach fire. Commercial abalone divers are not allowed to use scuba equipment—only amateurs. If you have skill with tools, the pretty shells can be carved into iridescent buttons, cuff links, and other jewelry.

Can our shoreline be saved? Not all of it. Many of our sandy beaches have already been destroyed. Half the estuaries along the California shore have been dredged, filled, or otherwise modified by human occupation. Hundreds of miles of seacoast are edged by coastal cities, which now must remain. And unknown numbers of ecosystems in coastal waters, marshes, and swamps have been degraded by chemical, oil, and sewage pollution. No amount of restoration is possible in some of these areas. The best solution is to preserve the relatively unmodified areas that still remain.

An examination of the ways in which our shoreline is being destroyed may provide some clues as to how we may preserve remaining natural areas for the future. We already know that beaches are destroyed by breakwaters, jetties, and dredging to satisfy the monetary desires of exploiters, marine promoters, ambitious coastal cities, and the demands of industries that insist they must be near the sea.

Environmental destruction results from many construction activities. Land clearing and housing construction leads to freshwater dilution of estuaries from rapid runoff. A single disturbance, such as dredging, can lead to topographic change, disturbance of tides and currents, deterioration of water quality, and decrease in recreation and commercial fishery. Increase in water turbidity can change shellfish and bird habitats and decrease the pollution cleansing capabilities of marshes and swamps. Residential construction prevents access for the seashore visitor. It also disturbs the breeding areas used by seashore wildlife. And finally, it affects the aesthetics of the seashore. John Hay and Peter Farb in *The Atlantic Shore* have written, "There is hardly a square foot of land along the North Atlantic coast, or for that matter in America, where the effects of technology, sometimes controlled, but often close to vandalism, cannot be observed."

Along Miami Beach, the cost of maintaining the sand supply is estimated at $1 million per year, with federal taxpayers footing 50 percent of the bill. Who reaps the monetary reward? You guess. Anne W. Simon, in her book *The Thin Edge: Coast and Man in Crisis,* tells us that lobster production is reduced by oil that spreads over their breeding grounds and that highly toxic PCBs dumped into the Hudson River estuary are moving up the food chain to human babies that drink their mothers' milk.

When Camp Pendleton was built during World War II, Oceanside, California, lost one of its finest beaches. The cost of replacing the sand was $800,000. Wave action is now taking it away. At Camp Del Mar boat basin, jetties interrupted the sea currents so that beach sands are no longer being replenished naturally by the sea.

In southwest Florida, the Deltona Corporation's Marco Island project sold lots for houses before the lots existed—they were underwater. The plan was to dredge and fill the mangrove swamps and hundreds of acres of grass-covered bottom. The mangroves are a buffer against erosion caused by violent storms, and the sea grasses are vital nursery areas for fish, shrimps, and other marine animals. Building permits have been denied, at least temporarily, by the Army Corps of Engineers. But there is no satisfactory solution yet.

Protection and restoration of America's shoreline is everyone's concern. Congress and a few states have begun to enact legislation. Plans and decisions for action are to be made under the Coastal Zone Management Act of 1972 and under more recent amendments which strengthen the program by improving and extending the management process.

Increasing public pressure is resulting in stricter zoning laws and antipollution regulations that restrain reckless development. Some progress has been made by placing certain shorelines under public control so their beauty and usefulness to the general public will be preserved. Cape Hatteras National Seashore and Padre Island National Seashore are examples of how exploitation can be reduced or controlled. Oregon has built state parks along the shore and excluded motor vehicles from certain sandy beaches in its effort to restrain degradation of the beaches.

The California Coastal Plan is a comprehensive approach to seashore problems with the purpose of perpetually safeguarding what remains of the undestroyed coastal resources. The plan identifies a four-mile strip of shoreline as the coastal Resources Management Area. It is supposed to provide free access to the ocean for the public, facelift of urban areas, and siting of power plants away from the sea and from geological fault zones. It has been violently opposed by the California Chamber of Commerce, building and construction trade unions, and the California Association of Realtors.

The Massachusetts Wetlands Protection Act is a good example of the complications that arise in administering the law and agreeing upon procedures to protect the coast. As pointed out in the Wetlands Project report, assembled by the Massachusetts Audubon Society, implementation of the act involves government officials, exploiters, contractors, consultants, students, special interest groups, and interested citizens. Some of the special interest groups have succeeded in getting loopholes into the law for the benefit of clients.

The major steps in implementing the act for any proposal that may affect wetlands involves determination if a site is subject to the act; notice of intent, and examination for completeness; review of the notice; public hearings; issuance of orders to continue or repeal the project; and appeal of orders, a process which involves many details of adjudicatory proceedings. Enforcement is the key to protection of wetlands, and its techniques involve an array of legal tools in and out of court. One may wonder if coastal plans will remain only a vision.

Ordinary citizens can and should participate in planning for coastal improvement and preservation. First, they should have an awareness of the problem and familiarize themselves with the principles of ecosystem management. Library research, consultation with professionals, and the obtaining of information from state and federal agencies, including environmental impact statements prepared for local areas of interest, are a beginning.

The National Audubon Society, the Sierra Club, and other national organizations with interests in environmental protection can supply factual information for citizens who wish to become better informed about coastal problems and issues. An invaluable list of the names and addresses of people and organizations at the state and national level is available in the annual *Conservation Directory* published by the National Wildlife Federation, 1412 16th Street, NW, Washington, D.C. 20036. An excellent summary of how citizens can participate is in *Who's Minding the Shore? A Citizen's Guide to Coastal Management*, available by writing to the Office of Coastal Environment, National Oceanic and Atmospheric Administration, Rockville, Maryland 20852.

The individual citizen can do little to influence the planning process of coastal management. He or she should consider joining or forming well-organized groups that can maximize participation in educating the public, lobbying, and informing planners, state and federal officials, and legislators of their views. Increased access of individuals and groups to administrative and judicial proceedings must be permitted if we are to save the remaining unexploited areas of our national seashores.

Fences protect the seashore by stopping the sand and allowing sea grasses to grow and bind the soil against the winds. This is at the edge of the Delaware–Maryland Shore.

11

Seashore Treasures

THE BORDER OF THE SEA HAS MANY TREASURES. There is challenge in the movement of the waves, antagonism when storms lash the dark waters into foaming fury, and mystery in the gentle shifting effects of light and the condition of the sky. Here, as the tides move back and forth, animate beings, such as sand crabs, sanderlings, and gulls, excite you with their strangeness, their beauty, and their uniqueness. If you observe closely, there are patterns of vegetable life, habitats for microscopic creatures, breakers in the moonlight, and dramatic colors in landscapes that never remain the same. The realization of such immense and awe-inspiring diversity is the beginning of love for the seashore.

Seashores vary a great deal and have appeal for almost everyone. Each shore has a different vista, different waves and waters, different dunes, marshes, estuaries, cliffs, and sandy beaches. Even the sand designs are unusual and vary from day to day and even hour to hour. Pebble marks, ripple marks, circles made by wind-blown blades of grass, bird tracks, bellybutton dimples made by clam siphons, and pellet patterns heaved up by marine worms give wide evidence of movement and vitality at the edge of the sea.

One of the treasures of the seashore is to be there and experience its environment through the senses. With bare feet you can feel the soft warm sand of the upper shore or leave footprints at the edge of the tide and see them vanish with the next wave. You can hear the cry of the gull over the sound of rolling stones and pebbles being tumbled smooth as the waves wash back to the sea. You can pick up shells, stones and driftwood, exploring the various textures with your fingertips. You can smell the distinctive odor of clean marshes or breathe in the almost tangible fog that drifts in from the sea when storms are brewing. Or you can sit on a high point and watch and listen and catch the mood of the sea while you enjoy the serenity of being alone.

It is easier to be alone on the beach in northern winters when the frigid winds blow and walking is more comfortable on the virgin frozen sand. There is less debris on the strand and the dunes are frosted or covered with snow. Animal tracks are evident, the snowy owl down from the Arctic is perched on a frozen mound, and the empty nest of pied-billed grebe still retains its mixture of mud and reeds.

These are only a few of the things you can see and do while exploring the seashore. For some people, painting at the beach is a joy. The mystery, beauty, and shifting combinations of light, clouds, and waves in a seascape challenge the artist's palette and brushes. The choice of scenes is not to be taken lightly—stormy weather, mares' tails and the sun's rays in the sky, and the composition and color of rocks and the sand can produce dramatic scenic effects if one can only capture on canvas the mood of the moment.

The constantly changing configuration of the seashore and the influence of the tides provide new grist for the photographer and the painter. A gentle sloping bottom near shore slows down the waves and makes them curl over into breakers with whitecaps that topple into foam. The gulls and cormorants then fly close to the sea. If the wind is gentle, the diving ducks disappear into the rollers and surface in the relatively calm troughs between waves. When the wind is strong, the gulls follow the air currents high above the sea around cliffs and prominences. At high tide, when the waves break far up the walls of cliffs or engulf house-sized boulders or barnacle-covered rock piles, the artist has the dynamics of nature in its grandest moments for subject matter.

Some seashore wanderers make simple sketches to enhance their nature notes. Birds, flowers, seashore mammals, dune forms, shipwrecks, and beach-strewn logs are likely subjects. I have derived pleasure from sketch maps of places visited from time to time or even years apart. These sketches illustrate the dynamics of nature when a repeat visit to a certain place reveals the disappearance of a

Snowy owls migrate from the Arctic in winter. They may be seen perched on dunes along Atlantic and Pacific shores.

sandy beach, the succession of plants in a seaside pond, or the migration of dunes and the uncovering of graveyards of trees formerly killed by drifting sand.

Memorable sights and experiences along the seashore are best caught and preserved by the quick action-stopping eye of the camera lens. The subjects are innumerable: cliffs, rock outcrops, sand patterns, clouds, valleys, and waves: moving things such as birds, animals, and plants swaying in the wind; and people wading in the surf, building sand castles, cooking meals on fires made of driftwood, or just throwing Frisbees over the beach.

There are joyful moments to be recorded on film. What angler is not happy to be photographed with a fabulous fish just hauled in from the sea? There are also poignant scenes to be recorded at the shore—a bird victimized by an oil spill at sea, an injured seal tossed up on the shore by storm waves, a shore littered with lumber and flotsam from a capsized ship. One of my pictures shows a baby's shoe, partly covered with sand and grayed and frosted by salt spray, amid the dune grasses. I often wonder who the baby was, if he or she were lost, or if the mother searched diligently for the little shoe? I left it there in the sand, but the photograph intrigues me still.

For close-up pictures of the sea animals and plants, you need a good camera, possibly one with macro lenses and bellows extensions. Your subject matter ranges from sand dollars, sea stars, and sand fleas to barnacles, limpets, and rockbound periwinkles. If you want to record sea lettuce plumes in tidewater channels, crabs in shallow water, or sea anemones in tide pools, lens filters of various kinds can enhance the color rendition on your film or eliminate reflections from the water surface. Electronic flashes can also illuminate crabs and other animals that like to hide in grottoes, open caves, and in the shadow of rocks.

In addition to the artists and photographers, there are other kinds of specialists among shore walkers and beachcombers. I admire them for their knowledge, their persistence, and their ability to find treasures along the seashores. There are even people who seek metal treasures with Geiger counters and mine detectors. Occasionally, they find an ancient encrusted sword or a few pieces-of-eight from Spanish explorers. There are specialists who look for nothing but rocks and semiprecious stones. The most avid collectors are those who pick up seashells, so abundant on many of our shores.

Collectors who get caught in the "shell game" are likely to develop a permanent stoop or a bend in the neck since they never look anywhere but down as they cruise the seashore. These shore walkers often present a ludicrous appearance. They wear old clothes, rubber boots, tennis shoes to preserve their feet, or bathing suits if they plan to invade the surf for the rarer shell specimens. Their accouterments include buckets, canvas bags, glass jars, bottles, and knives for prying mollusks from rocks. I have seen women wearing pack sacks in front, covered with raincoats, the whole of which gave them the appearance of advanced pregnancy. However they appear, and wherever you see them, digging, wading, diving in the ocean, slogging through mud flats, or clambering over slippery rocks, do not laugh. Remember that they are dedicated people who truly love the seashore and one of its greatest treasures.

The shell collector soon learns that seashells occur in a multitude of habitats. It helps if one knows something of the life histories and the ecological relationships of animals that make shells. This increases the possibility of collecting a wider variety of shells and finding some of astounding beauty. For the dedicated collector, the desire for identification soon arises, and he finds that most shells fall into groups.

The commonest group is the bivalves, with two-part shells, such as clams and oysters. Then there are the univalves, with only one shell, usually of spiral shape, including snails, abalones, whelks, and conches. Tusk shells occur in the form of tubes. These elongated shells belong in a separate class of the mollusk group. Chitons belong to still another class of primitive mollusks with a shell consisting of eight plates, which fall apart when the girdle that binds them together is removed. These are especially hard to preserve as specimens and are generally disregarded by collectors, as are the cephalopods.

Where to collect shells along the seashore presents few problems. The entire shoreline of the United States offers a variety of mollusk shells for the collector. Many species of clams are abundant on both the Atlantic and Pacific coasts. The latter coast also offers a large variety of chitons, limpets, and sand dollars. Mussels are superabundant on some rocky outcrops, where they may be gathered at low tide. Skin divers in California hunt abalones, prying them from rocks with an iron bar similar to a tire iron. A license is required to collect abalones.

The Gulf coast, from Florida to Texas, offers many colorful and spectacular shells which can be easily collected along the shore. Some of the most colorful shells in North America are found along the east coast of Mexico and along the west coast of Florida. The latter coast offers Scotch bonnets, Fargo's worm shell, the common purple sea snail, the spotted slipper, the king's crown, and the lightning whelk. Sanibel Island, off the coast from Fort Myers, is a collector's paradise; more than 300 species have been sighted here, including calico scallops, turkey wings, junonias, murexes, paper figs, and rose tellins. The Florida Keys, as almost every visitor knows, are famous for the pink-hued queen conch. Also present are cowries, jewel boxes, helmet shells, and bleeding teeth.

Limpets, which resemble Chinese hats, some with a hole in the top, occur in Florida and along both ocean coasts. These interesting animals attach themselves to rocks by means of broad feet, which pull the shell tightly against its anchorage. They graze over their rock for algae but always return to their same resting spot. They can be collected by inserting a knife blade between the foot and the rock when the animal lifts its shell to move or breathe. The limpets with holes in the tops of their shells are called keyhole limpets. Some species attach themselves to seaweeds instead of rocks.

The experienced collector knows that the open beaches, the tidal flats, the subterranean world of sandy areas and muddy bottoms, the wave-washed rocks, and the shoreward dunes are the realm of shells. Algae, especially the kelps, washed up from the sea are also collecting sites for small shells. Do not neglect the undersides of boulders for handsome shell specimens. But return the rock to its original resting place so other creatures can resume their living habits. Even the surfaces of other undersea animals can be habitats for shells, as are waterlogged pieces of wood which are homes for wood-boring clams. Last of all, the stomachs of bottom-feeding fish may contain specimens you will never find in shore habitats.

Some collectors use wire strainers when collecting tiny mollusks from sand or mud. Boxes with wire screen bottoms of different meshes can turn up a surprising number of shells so tiny that a hand lens or microscope is almost a necessity for identification. Seaweeds rapidly dunked in a bucket of seawater may also result in a catch of numerous specimens. A face mask and snorkel will help you search for specimens in shallow water.

Cleaning shells, once they are collected, can be simple or elaborate and frustrating, depending on the species. Small snails and clams can be placed in cool water, brought to a boil for 10 minutes, and then cooled slowly to prevent cracking the shell. The meat then can be pulled out with a hairpin, bent wire, or dental probe. All experienced collectors stress the need to preserve the operculum, or "trap door," that some snails have. The cleaned shells should be dried in shade, never in sunlight.

Special cleaning methods should be used on shells to avoid destroying their high polish and their periostracum, or outer protein layer. Oiling after cleaning is required for many seashells to preserve their color and patterns. Mineral oil is better than olive oil for this purpose. An excellent book, a must for the serious shell collector, is *How to Clean Seashells,* by Eugene Bergeron. The author debunks some of the methods used by amateurs who destroy the natural beauty of many shells by boiling them to remove their soft inner parts. For certain shells, refrigeration, alcohol, formalin, lye, muriatic acid, and even bacteria are better than boiling. Certain univalves, bivalves, and chitons require special methods. The last-named must literally be bound to glass and tied down before preservatives are applied.

I learned my first lesson about removing hermit crabs from their shells some years ago after collecting some fine large specimens at

Alligator Point near Tallahassee, Florida. Boiling was out of the question since that forces the crab to retreat to the inside of the shell before it succumbs to the killing heat. The stink that follows this procedure is terrible and long lasting. Mrs. P.F.W. Prater, a great friend who lived in Tallahassee, would not allow me inside the house with them. Later, I solved the problem by holding the rear ends of the shells against the hot motor of my car. As the shells became hot, the crabs popped right out. The simplest method of all, however, is to place the shells in fresh water and add a small amount of chlorine bleach. The crabs then crawl out.

One more task remains for the serious shell collector: to label the shells with numbers written with India ink and record the number and other pertinent information in a notebook. The label can be kept with the shell, though it is apt to be lost or separated from the shell. The label should at least include the geographic location and habitat where the shell was found, the date collected, the scientific and common name if these are known, and the name of the collec-

Skin divers go to sea on hand-propelled boats and dive in shallow water to spear a catch of edible fish. The wet suit keeps them warm in cold ocean water, and the flippers enable them to swim with considerable speed.

tor. If they are small, store the shells in boxes, drawers, and in closed containers away from the sunlight. If the specimens are large, display them in glass mounts or on shelves. Trade them if you wish. Whatever you do with them, they will always bring back pleasant memories of the seashore.

Another form of collecting likely to give the beachcomber a permanent stoop is rockhounding. The rockhound haunts the low-tide zone and the cliff bases, scrutinizing every pebble and stone in search of agates, jasper, petrified wood, and other stones and minerals. These explorers of the seaside are impervious to the weather and oblivious to fog, rain, or wind. They love the winter months when storms bring waves that scour the sand from gravel beds or expose stones that have broken from bluffs and cliffs and fallen into the sea.

The New England rock collector does not find the variety and abundance of desirable rocks that are available to Pacific coast rockhounds. East coast rocks are mostly granite, scraped bare by former glaciers. Some igneous rocks and metamorphic rocks contain intrusions or veins of minerals, which make collector's items. Limestones and red sandstones also may be added to the collector's cabinet. The Gulf shore and the sandy beaches from Florida to Cape Cod offer little to the rockhound. These beaches are the realm of the shell collectors.

Pacific shores open up a world of variety for rock and mineral collectors. California beaches present an interesting selection of stones. Agate, petrified wood, and jade occur on the beaches near Crescent City. Petrified wood is found on Jalama Beach. The seaside at Bodega Bay, Drakes Bay, and Duxbury Point in San Francisco County offer chalcedony pebbles and petrified whalebone. Rhodonite occurs at Lime Kiln Creek Beach near Lucia in Monterey County.

If you drive southward down the west coast of Baja California, you will be in territory virtually unprospected by rock collectors. By exploring not only the beaches but some of the nearby islands, you may find petrified wood and agates. At the curio shops, you can get bargains in prehnite, selinite crystals, and jade-green actinolite if you're willing to haggle a bit with the store owners.

Rockhounding on Northwest beaches produces many desirable stones. When stormy weather exposes the best material from gravel beds, hundreds of stone seekers scour the beaches for rhyolite rocks and basalt containing agate nodules. Agates are found from Eureka, California, to the Straits of Juan de Fuca. Other stones are petrified wood, serpentine, and a grossularite garnet referred to as Oregon Jade. Agates occur in such varieties as carnelion, ribbon, cloud, moss, and rainbow.

One of my favorite collecting areas is the Tillamook County beaches. One never knows when the stones will be exposed, but Bayocean, north of Tillamook Lighthouse, is a good spot to look for white, pink, and green jade. The original town has been lost to winds and storms, but the cobble and pebble pavements at tide edge contain a variety of desirable stones. Colorful rocks suitable for tumbling and polishing also are found on gravel banks near Westport, Washington, and along the shores of Vancouver Island, British Columbia.

One of the greatest finds the beachcomber can make on Pacific shores is a Japanese glass float. These hollow glass balls sometimes spend years crossing the Pacific Ocean before landing on Vancouver Island or the Washington or Oregon coast. They break loose from fishing nets in the vast stretches of the sea and then ride the waves until storms and high winds finally bring them to land. Some people spend years looking for them and never find a single float. Other beachcombers hit the jackpot and come home with dozens, sometimes with pieces of fishing nets still attached.

My wife, on our numerous visits to Pacific shores over the course of years, searched and never found a single one. Then one morning, while looking out the window of a friend's house high above the beach, she saw a large float coming in with the tide. It was five o'clock in the morning. She was in her robe and I in my pajamas. There was no time to dress. I dashed madly down the cliff road, across the beach, and into the waves to retrieve the float, which was more than a foot in diameter. I was just in time because three other men were running to catch the beautiful greenish ball. Beachcombers, I realized, are eternally vigilant, and apparently they never sleep.

Most of the glass floats arrive on the northern Pacific coast via the Kuroshio current from Japan. These fascinating, delicately colored spheres sometimes come ashore by the hundreds. They are eagerly sought by amateurs and professionals alike. Some enthusiastic collectors cruise the beaches at night hoping to spot reflections of floats in the beams of their beach buggy headlights. Even more sophisticated collectors fly in planes and helicopters while they look for floats in driftwood, at the edges of cliffs, and in bays that are accessible only by boats.

Floats are not always spherical. Some are shaped like cylinders. Others are pear-shaped, and a few resemble spindles. They are made not only in Japan but in Russia, Korea, Germany, and the United States. Their methods of manufacture and their classification are thoroughly covered in *Beachcombing for Japanese Glass Floats,* by Amos L. Wood. His book includes an illustrated list of trademarks that are imprinted on the glass. Oddities include floats made with tubes through which a rope may be threaded and double-ball floats. Some of these vagabonds of the sea have been in the water so long they are covered with barnacles.

Most of the floats my friends have found are grapefruit-sized or slightly larger. One of the largest ever found was 58 inches in circumference. You have a real prize if you find one 20 inches in diameter or even one that is basketball-sized. Colors vary from clear glass to milky white, green, red, olive, and even gold, depending on the minerals that have been added to the glass during manufacture. Some plastic floats are now being made.

It is well to be aware of imitation floats made for sale in curio shops along the coast. These are usually free of bubbles in the glass and they rarely carry a trademark even though they carry a label, *Made in Japan.* Instead of buying a float, your greatest thrill can come from finding one along Pacific shores. Look for them after storms, strong winds, and high tides, especially in late fall and early spring. Some will be lodged in driftwood at the upper tide line. Newport and Windchester Bay in Oregon are good hunting areas. Nehalem sand spit near Tillamook is said to generate a yearly average yield of 15 floats per mile per month. If you find one, try to identify the manufacturer and the meaning of the symbol with the aid of Amos L. Wood's book.

Clam digging along the seashore is not a sedentary occupation. It involves knowledge of clam habitats and habits. It requires expenditure of energy, especially if you go after the deep ones. It can be a sociable sport or an individual endeavor, and it provides some delectable items for the dinner table.

Clams are one of the most numerous edible animals of the seashore. But you must know where to find them, which are the most edible, and how to collect them. Some are easy to collect. The blue mussel (*Mytilus edulis*), on rocky shores from Maine to Connecticut, makes a gourmet dish when steamed and then dipped in butter or vinegar. They are easily found, but you must pull them loose by breaking the byssus, or threadlike attachment, that holds them to the rocks.

Surf clams (*Spisula*) are large clams, from 6 to 7 inches long. They make excellent chowder. They are not obtainable on the shore since they live at depths of 40 to 200 feet. At Truro on Cape Cod they are sometimes washed up on the beaches by the thousands after severe storms.

Even when the beach seems deserted, the clams are there. You can spot their locations by observing jets of water that squirt from the sand as you walk along the shore. The common squirting clam is the soft-shell clam (also called steamer clam or long-neck clam). It squirts from its muscular double siphon when disturbed. It is easy to dig since its foot remains stationary and does not burrow rapidly

These clam diggers use garden rakes to uncover the shallow burrowing clams at low tide on Tillamook Bay, Oregon. The bay bottom literally gets plowed, but the clam population is not noticeably diminished. Common bay clams along the coast include: cockles, collected with rakes to depths of 4 inches; horse or gaper clams, dug with shovels to depths of 12 to 18 inches; butter clams, 6 to 8 inches deep; and littleneck clams, either raked or dug with potato forks.

into the sand. When looking for these and other clams, keep an eye out for the round or wedge-shaped holes in the sand which may indicate the spot where a clam has withdrawn its siphons. Some of these bellybutton marks, however, may be caused by worms or even by air bubbles that have escaped to the surface of wet sand.

One particular clam, native to Long Island and New England, goes by different names, depending on its age and size. It is the hard-shelled clam, or quahog, which lives to the ripe old age of twenty to twenty-five years. In restaurants, these clams are listed on the menu as littlenecks or cherrystones. The young ones are eaten on the half shell. Older ones go into chowder. A good chowder clam is the quahog (*Mercenaria* sp.), which can be captured in bays from a boat. These clams are lifted from the muddy bottom with a two-handled rake that works like a pair of tweezers. The giant quahog (*Artica* sp.) is difficult to open, but the task can be accomplished by warming it in an oven until the shell opens slightly. When cool, a knife can be used to cut the muscle that holds the shells together.

The razor clam of eastern shores, elongated with a short siphon and a big foot, is a tricky customer. Like its relative on Pacific shores, it digs rapidly and deep. If you are not quick with the shovel, you lose the clam and end up in sandy muck up to your elbow, still trying to grab the creature. Another popular New England soft-shelled clam, the Ipswich clam (*Mya* sp.), is delicious when fried or steamed.

The Pacific coast has its share of clams, and the methods for catching them reveal considerable ingenuity on the part of clammers. Any implement used to capture clams is called a clam gun. These include specially shaped shovels, rakes, hooks, darts, stovepipes, and galvanized cylinders with open bottoms and closed tops. The last have a thumbhole in the top which is left open as the "gun" is centered over a sand dimple and pushed down a foot or more. The thumb is then placed over the thumbhole to create suction and the clam is lifted to the surface, along with several quarts of sand. It is strictly an instrument for the amateur. I find it much more fun to insert the blade of a long handled shovel on the landward side of a razor clam, lift the sand quickly, and grab instantly before the critter starts digging toward China.

The most expert razor-clam digger I have ever known was my friend Larry Mays, who used to dig clams commercially. There is a strict bag limit for amateur diggers. Larry would wade into the shallow surf and spot clams by seeing their siphons exposed in clear water. His self-imposed quota was to dig three clams between waves. Since waves come every seven to ten seconds, that meant eighteen to twenty clams per minute. His record was several hundred dozen clams in one tide. My record was about twelve clams in two hours of digging.

The razor clam's foot is an intriguing organ. The beast lives with its siphons pointing up and its foot pointing down and its digging ability is remarkable. A friend of mine once emptied a bucket of clams on the sand and went to his car to get a sack. When he returned, all his clams had disappeared. They simply dug in by extending their feet into the sand at a right angle and raising their shells to a vertical position. Once started into the sand, they expanded their feet to serve as anchors while they pulled their shells downward. They were soon out of sight.

West coast clams are of many varieties. The Washington clam reaches commercial size when it is about 1.5 inches in diameter, but it grows to a length of 8 or 9 inches. Gaper clams or horse clams, found in bay mud, weigh up to 4 pounds. Their meat is tough and must be chopped fine. Pismo clams are found in sand under the breaker line from Monterey to Baja California. They are dug with potato forks. Rock cockle, or littleneck clams, occur on gravel bars and reefs near bays and inlets from California to Washington. They must be 1.5 inches in diameter to be taken legally.

Littleneck clams are easily dug with a garden trowel. Washington clams are speared with a heavy rod 3 feet long with a 2-inch right-angle hook at the end. The procedure is to locate the hole, push the rod down, turn the rod under the clam, and lift. Pismo clams may be located by the tufts of hydroids which grow at the end of the siphons. The giant of them all, the geoduck, squirts a goodly stream of water on mud flats, and weighs several pounds. It lives several feet below the surface and requires herculean effort before it comes into the collector's hands. A large stovepipe or piece of small culvert material is gradually lowered toward the clam. This metal shield keeps mud and debris from caving in as the digging proceeds. The excavation sometimes leaves a hole in which one could bury a small cookstove.

There are many recipes for preparing clams for the table. Tomato clam chowder and chowder with chicken broth are delectable. Clams can be baked with corn or roasted on the half shell. They can also be stuffed with water chestnuts, fresh ginger, and green onions combined with lean pork or beef. And, of course, they can simply be steamed or eaten raw on the halfshell.

I have dug many clams, but to my great regret I cannot eat them. When I lived in Portland, Oregon, I used to eat the best clam chowder in the West at a restaurant on Broadway Avenue. By keeping a diary record of my migraine headaches and temporary blindness, it appeared that I am allergic to clams. After that self-diagnosis, I never ate clams in any form for two years. Then on a trip from Reno to Portland I stopped for lunch at the old hotel in Lakeview, Oregon. The clam chowder looked and smelled wonderful. So I tried it. I made it to the Horse Ranch near Fort Rock, Oregon, and then crawled on hands and knees in the sagebrush, blind and sick all over, for more than two hours. Clams are fun to dig and wonderful food for those who can eat them. Success to the happy clam eater.

Fishing is a sport enjoyed by millions of people who visit the seashore. Fishing can be enjoyed almost anywhere because there are so many kinds of fish. In New England, the herring run crowds the stream bank with people. The smelt run in Oregon and Washington becomes an extravaganza of citizens of all ages carrying all types of equipment. The grunion run in California evolves into a midnight spectacle as bare-handed fishermen indulge in what is almost a tribal rite. But these forays are only seasonal events. The real fishing at the edge of the sea is an exercise in patience, knowledge, and skill since a different approach is necessary to catch each particular kind of fish.

Rockfish and scorpionfish in California waters are represented by more than 50 species. Most of these are highly desirable for food and many are used for food by larger fish. The adult forms of many are found only in relatively shallow water, often close to the shore and readily accessible to surf casters or charter-boat fishermen, who catch them by the millions.

The Florida Department of Natural Resources estimates there are 600 species of fish in the state's coastal waters. Fishing in the Keys can make you believe this claim. The problem for anglers who come to the seashore from inland states is to get acquainted with fishing methods, to learn where the best spots are, and to choose the right bait for certain kinds of fish. It is best to seek advice from local fishermen.

You can fish from piers and bridges in some places. Charter boats are available for most visitors and the skippers know where to fish, so success is almost guaranteed. Most seashore visitors will not be involved with ocean game fish such as blue marlin, sailfish, amberjack, barracuda, cobia, dolphin, and king mackerel. But tarpon can be caught in ditches that flow to coastal waters. Monsters weighing over 120 pounds can be caught with trolling spoons and giant plugs.

The Gulf of Mexico offers fascinating fishing on inshore waters. Croakers are bottom-fished with shrimp or cut bait in bays, from bridges, and in the surf. Bonefish are light-tackle speedsters and are interesting because they can be seen feeding on shallow flats. For them, live shrimp make good bait. The cast should be made to one side of fish that are feeding on the bottom, their tails waving above the surface. You will be glad if you have 200 yards of backing on your reel when a fish starts its long run.

If you have the opportunity to fish in the mangrove forests, or in creeks and bays in the Ten Thousand Islands, snook will be there and provide top sport. Other inshore fish are mullet, speckled trout, sheepshead, Gulf whiting, and Atlantic threadfin. If you intend to indulge in the fabulous saltwater fishing in Florida, you should get a copy of the *Outdoor Guide,* published by the *Miami Herald.* The "Fishing Guide," as it is called, describes many methods of fishing

Fishing for sea perch. These little beauties come into rocky areas at high tide. Sea worms make good bait, and a fish may be caught on nearly every cast.

and boating, gives tide tables, and illustrates dozens of fish in color.

North of Cape Hatteras, the fish fauna becomes quite different from that of the south Atlantic and Gulf regions. From Maine to Cape Cod, for example, some of the bottom fishes are halibut, winter flounder, haddock, and hake. Many of these are taken commercially. But amateur anglers find much sport in estuaries, which are nursery areas for smelt and tomcod, and the permanent residences of winter flounder. Anadromous fish also pass through estuaries from the sea to spawn in local rivers. These migrants include the Atlantic salmon, shad, alewife, and blueback herring.

Striped bass are frequently caught by trolling from boats or casting from shore with a variety of lures that range from eels, clams, and mackerel to plugs, spoons, metal jigs, and feathers. On sand or rock bottoms, porgies occur near shore during warm months. They seldom weigh more than 2 pounds but can be bottom-fished from shore with clams, mussels, shrimps, killifish, small spoons, or spinners.

If you are unfamiliar with Pacific Coast fishing, you should obtain a copy of *Anglers' Guide to the United States Pacific Coast*, by James L. Squire, Jr., and Susan E. Smith. The great abundance of inshore fishes, as well as marine species, offers endless variety for the sport angler. You do not have to fish offshore for salmon, albacore, tuna, or marlin to enjoy angling. Instead, there is pier fishing, rock fishing, surf fishing, private boat fishing, and charter-boat fishing up and down the coast. The *Anglers' Guide* lists basic criteria for the fisherman who wishes to try his or her luck for anything from halibut, lingcod, ocean whitefish, perch, sand sole, and striped bass to bottom fish.

The rugged Oregon–Washington coast with its sandy beaches, bays, inlets, and estuaries between headlands offers some of the most interesting fishing anywhere. Salmon are one of the primary attractions for anglers from California northward to British Columbia. The major species on the open coast are Chinook and coho salmon, sea-run cutthroat, and Dolly Varden trout. Bottom fish, such as rockfish, cabezon, lingcod, and greenling, make excellent catches for shoreside anglers who fish from jetties or from boats on bays.

There is a kind of beach exploration that has no relation to the scavenging, plundering, and denudation of shore life as practiced by many beachcombers. I do not refer to the wholesale destruction of shorelines, bays, and estuaries by exploiters, power stations, junk housing at the water's edge or the pollution from industry. Instead, I refer to our individual responsibility to protect our natural heritage by abandoning greed, forgoing collection of living things only to throw them away, and preserving the good environment that still remains along some of our seashores.

Each of us can practice self-restraint when we visit various seashore habitats. Actually, there can be more pleasure in observing the biological characteristics and activities of animals, especially the rare ones, than catching, digging, and collecting more than we want or need. Unsupervised children can be offenders when given the freedom to carry off whole baskets full of starfish, which will surely die and create a stench because the youngsters do not know how to preserve them.

Among the worst offenders are biology classes from schools. I experienced this once when a college group from an eastern university came to Colorado and invaded one of our protected experimental areas where native plants and animals had been allowed to live without disturbance for more than fifteen years. The students literally "went to ground" and dug up the place until it had the appearance of a plowed field. They excavated rare plants, denuded the bark from trees in search of boring beetles, and appropriated bird nests with eggs in them.

Biology classes commit similar offenses along the seashore. Each student is inclined to take one or more specimens of a given species when one would be sufficient for class demonstration. Eugene N. Kozloff, in *Seashore Life of Puget Sound, the Strait of Georgia, and the San Juan Archipelago,* has offered some rules for those who would enjoy seashore recreation without destroying the heritage for other people yet to come.

Kozloff suggests the use of flora and fauna from floating docks and pilings to remove the pressure of collecting from natural areas. Avoid collecting from natural areas. Avoid collecting rare species that are found only in limited localities. Make good use of what you do collect. Replace rocks after you have examined their undersides and fill holes when you dig sea animals. Know and obey the fish and game laws of the state and observe the rules for collecting or not collecting in designated natural areas.

Biological excursions and plain nature study can be rewarding even if not a single living animal is disturbed or collected. Wholesale destruction of habitat brings many animals to extinction or reduces their presence for the enjoyment of others. Some fishermen, curio collectors, and divers are none too virtuous in this respect. They want the biggest fish, the largest octopus, or the rarest sea turtle to display to so-called admiring friends. Happily, most scarce animals are protected by law or the numbers that can be taken legally are restricted. This even applies to clams and crabs.

You do not have to be a trained biologist to have an interesting and instructive visit to the seashore. There are books for most of the seashores of America, written about specific areas, which will help you to identify the creatures of the tide pools, the beach, and the upper shore. These books offer suggestions on where to look and how to observe and understand the activities and habits of everything from crabs, starfish, barnacles, and jellyfish to sea lions and whales. The memories of seeing these creatures will persist long after the day you watched them in their true homes.

Even a small tide pool can be an instructive place. If hermit crabs are present, by remaining motionless, you may see them change or borrow a shell. If the shell is empty, the crab will test it with claws and feet to make sure no creature is inside. You may even see a fight between crabs over a favorite shell. If the pool at first seems lifeless, wait and watch. The bumps on rocks may turn into limpets or snails. The barnacles and sea worms may open their plumes to catch the soup of the sea. The sea anemone may open like a colorful flower. Then you may become aware of an almost transparent shrimp or a delicate nudibranch walking upside down on the water surface film. Crabs may make their appearance, or a keyhole limpet may spread its broad foot and move along the bottom. And then that most colorful of goldfish creatures, the orange garibaldi, may swim into view.

All this you can see without moving. And do not fail to observe the colorful display of seaweeds which add texture, hue and hiding places for creatures of the pool. The seaweeds, especially the algae, fascinate some of the more venturesome people who love the sea. There is some danger in collecting algae, especially if one goes into deep water where the most fabulous of these plants are found. You can slip and be immersed if you tread on slick rocks. The scuba diver, of course, can collect in the algal forests beyond the low tide zone where many species are never exposed to air. There is also good collecting in the intertidal zone, where many algae cling to rocks or lie prostrate on the shore. The big seaweeds also become available when storm waves tear them from their moorings and bring them to shore. Do not forget that small epiphytic forms of algae grow on the larger species.

This book is not the place to discuss algae collecting in detail. The expert goes equipped with cloth or plastic sacks, buckets, small vials or bottles, knives for scraping small specimens off rocks, and chisels and hammers for collecting stony algae. He also knows that many algae lose their color and that some must be preserved quickly before they putrefy. Commercial formalin, 3 to 4 percent, neutralized with borax in seawater is a better preservative than alcohol, which has a bleaching effect.

In warm climates, some specimens can be dried in the field. Kelps will become leathery as they dry in the sun. Holdfasts should be included for identification purposes. Some fragile and feathery specimens can be floated onto herbarium paper, which is then covered with waxed sheets and placed between blotters. Blotters should be changed frequently.

When you collect specimens, be sure to label them as to date and time of collection, tide level, stratum, general habitat, and name of collector. Then at some future time go to the books to identify what you have, to learn its life history, if known, and to renew pleasant memories of days spent on the seashore.

There are times when I find it a pleasure to go to the seashore alone where I can sit on a high dune, or a promontory above the beach, in a place of solitude. Here, away from the human labyrinth, life comes to a standstill, at least for an hour. I look at the waves far below and I see living things carrying on their activities. The sanderlings are moving in and out with the waves. The gulls are standing quietly on the sand, preening their feathers or just looking. A beachwalker passes far below. Not once does he look back or upward. He looks only at the sand or the sea.

A dead seal lies on the beach. A gull pecks at it, but the hide is too tough and the gull walks away. The dead seal seems to have taken possession of the landscape. I think of the food chain that enabled it to be born, grow, live, and die. The chain is still at work. The seal, by virtue of its meat and bones, will become a part of it again.

Far down the beach I see chunks of driftwood, bleached gray by the sun and buffed smooth by wave-washed sand. They seem larger

Gulls and other scavengers eat whatever dead thing comes to shore, be it a crab, a fish, or even a seal. This is part of nature's recycling system.

than they actually are. Henry David Thoreau observed this phenomenon years ago. And now I remember how the raised edges of my footprints in the mud of Tillamook Bay once looked like mountain ranges when I viewed them from the flats a few hundred yards away.

My foot dislodges a pebble and it rolls down toward the beach. It brings my attention to the near view. The waves will reach the pebble at high tide, or the next storm will toss it around, nibble off molecules of its surface, and eventually reduce it to grains of sand. The pebble makes me contemplate the bedrock of the world and the sea which covers much of it with its living creatures that have been carrying on for millions of years.

The thought occurs that they will be there in the years and millennia yet to come. But if and when I come to this place again, will the shore be the same? Will anything be quite the same, even if I do not return? Really, it does not matter. But I shall remember this hour of solitude.

Bibliography

ABBOTT, R. TUCKER. *How to Know the American Marine Shells.* New York: New American Library, 1970.

ALLYN, CHARLES F. *The Great Outdoors Book of Skin and Scuba Diving.* St. Petersburg, Fla.: Great Outdoors Publishing Company, 1971.

AMERICAN PETROLEUM INSTITUTE. *Saving Oiled Seabirds.* Washington, D.C.: 1978. (Distributed by American Petroleum Institute, Distribution Services, 2101 L Street, NW, Washington D.C. 20037.)

AMOS, WILLIAM H. *The Life of the Seashore.* New York: McGraw-Hill Book Company, 1966.

ANDREWS, JEAN. *Sea Shells of the Texas Coast.* Austin, Tex.: University of Texas Press, 1971.

———. *Shells and Shores of Texas.* Austin, Tex.: University of Texas Press, 1977.

BAKKER, ELNA. *An Island Called California.* Berkeley, Cal.: University of California Press, 1972.

BARNES, R.S.K. (ED.). *The Coastline: A Contribution to Our Understanding of Its Ecology and Physiography in Relation to Land-Use and Management and the Pressures to Which It Is Subject.* New York: Wiley-Interscience, 1977.

BERGERON, EUGENE. *How to Clean Seashells.* St. Petersburg, Fla.: Great Outdoors Publishing Company, 1971.

BERMAN, BRUCE D. *Encyclopedia of American Shipwrecks.* Boston: The Mariner Press, 1972.

BERRILL, N. J., and JACQUELYN BERRILL. *1001 Questions Answered About the Seashore.* New York: Dover Publications, 1976.

BIRD, E.C.F. *Coasts.* Cambridge, Mass.: MIT Press, 1969.

BROWN, VINSON. *Exploring Pacific Coast Tide Pools.* Healdsburg, Cal.: Naturegraph, 1966.

CAREFOOT, THOMAS. *Pacific Seashores: A Guide to Intertidal Ecology.* Seattle, Wash.: University of Washington Press, 1977.

CARSON, RACHEL.*The Edge of the Sea.* Boston: Houghton Mifflin Company, 1963.

CHAPMAN, V. J. *Coastal Vegetation.* 2nd ed. New York: Pergamon Press, 1976.

CLARK, JOHN. *Coastal Ecosystems.* Washington, D.C.: The Conservation Foundation, 1974.

COE, WESLEY ROSWELL. *Starfishes, Serpent Stars, Sea Urchins and Sea Cucumbers of the Northeast.* New York: Dover Publications, 1972.

COMMITTEE ON GOVERNMENT OPERATIONS. *Protecting America's Estuaries: The Potomac.* Forty-fourth Report by the Committee on Government Operations. 91st Congress, 2nd Session. House Report No. 91-1761. Washington, D.C.: December 16, 1970.

CONSERVATION FOUNDATION. *The Decline of Galveston Bay.* A Conservation Foundation Study, directed by James Noel Smith. Washington, D.C., 1972.

COOPER, WILLIAM S. *Coastal Sand Dunes of Oregon and Washington.* Memoir 72. New York: Geological Society of America, 1958.

CRISP, D. J. (ED.). *Grazing in Terrestrial and Marine Environments.* British Ecological Society Symposium Number Four. Oxford: Blackwell Scientific Publications, 1964.

CUSHING, D. H. *Marine Ecology and Fisheries.* New York: Cambridge University Press, 1975.

DANCE, PETER (ED.). *The Collector's Encyclopedia of Shells.* New York: McGraw-Hill Book Company, 1974.

DAUGHERTY, ANITA E. *Marine Mammals of California.* Sacramento, Cal.: California Department of Fish and Game, 1965.

DAVIS, L. V., and I. E. GRAY. "Zonal and Seasonal Distribution of Insects in North Carolina Salt Marshes." *Ecological Monographs,* 36:275–295 (1966).

DAWSON, E. YALE. *How to Know the Seaweeds.* Dubuque, Iowa: William C. Brown Company, 1956.

———. *Seashore Plants of Northern California.* California Natural History Guides No. 20. Berkeley, Cal.: University of California Press, 1966.

DeCOURSEY, P. J. (ED.). *Biological Rhythms in the Marine Environment.* Columbia, S.C.: University of South Carolina Press, 1976.

DOLAN, ROBERT, PAUL J. GODFREY, AND WILLIAM E. ODUM. "Man's Impact on the Barrier Islands of North Carolina." *American Scientist,* 61:152–162 (March–April, 1973).

DUCSIK, DENNIS W. *Shoreline for the Public.* Cambridge, Mass.: MIT Press, 1974.

DUDDINGTON, C. L. *Beginner's Guide to Seaweeds.* New York: Drake Publishers, 1971.

EDWARDS, PETER. *Illustrated Guide to the Seaweeds and Sea Grasses in the Vicinity of Port Aransas, Texas.* Austin, Tex.: University of Texas Press, 1976.

GANTZ, CHARLOTTE ORR. *A Naturalist in Southern Florida.* Coral Gables, Fla.: University of Miami Press, 1971.

GATES, DAVID ALAN. *Seasons of the Salt Marsh.* Greenwich, Conn.: The Chatham Press, 1975.

GODIN, GABRIEL. *The Analysis of Tides.* Buffalo, N.Y.: The University of Toronto Press, 1972.

GOSNER, KENNETH L. *A Field Guide to the Atlantic Seashore: Invertebrates and Seaweeds of the Atlantic Coast to Cape Hatteras.* Boston: Houghton Mifflin Company, 1979.

————. *Guide to Identification of Marine and Estuarine Invertebrates: Cape Hatteras to the Bay of Fundy.* New York: Wiley, 1971.

GOTSHALL, DANIEL W. *Fishwatchers' Guide to the Inshore Fishes of the Pacific Coast.* Monterey, Cal.: Sea Challengers, 1977.

GOTTO, R. V. *Marine Animals: Partnerships and Other Associations.* New York: American Elsevier Publishing Company, 1969.

GOWANLOCH, JAMES NELSON. *Fishes and Fishing in Louisiana.* Baton Rouge, La.: Claitor's Book Store, 1965.

GREEN, J. *The Biology of Estuarine Animals.* Seattle, Wash.: University of Washington Press, 1968.

GUBERLET, MURIEL L. *Sea Weeds at Ebb Tide.* Seattle, Wash.: University of Washington Press, 1956.

HAY, JOHN, AND PETER FARB. *The Atlantic Shore.* New York: Harper & Row, Publishers, 1966.

HEDGPETH, JOEL W. *Introduction to Seashore Life of the San Francisco Bay Region and the Coast of Northern California.* Berkeley, Cal.: University of California Press, 1962.

HILLSON, C. J. *Seaweeds: A Color-coded, Illustrated Guide to Common Marine Plants of the East Coast of the United States.* University Park, Pa.: Pennsylvania State University Press, 1977.

HINTON, SAM. *Seashore Life of Southern California.* California Natural History Guides, No. 26. Berkeley, Cal.: University of California Press, 1969.

HOESE, H. DICKSON, AND RICHARD H. MOORE. *Fishes of the Gulf of Mexico, Texas, Louisiana, and Adjacent Waters.* College Station, Tex.: Texas A & M University Press, 1977.

HOFFMEISTER, JOHN EDWARD. *Land from the Sea: The Geologic Story of South Florida*. Coral Gables, Fla.: University of Miami Press, 1974.

IDYLL, CLARENCE P. "Grunion, the Fish That Spawns on Land." *National Geographic:* Vol. 135, No. 5 (May 1969).

JACOBSON, MORRIS K., AND WILLIAM K. EMERSON. *Shells from Cape Cod to Cape May*. New York: Dover Publications, 1971.

KAUFMAN, WALLACE, AND ORRIN PILKEY. *The Beaches Are Moving*. Boston: Houghton Mifflin Company, 1979.

KEATING, BERN. *The Gulf of Mexico*. New York: The Viking Press, 1972.

KELLEY, DON GREAME. *Edge of a Continent: The Pacific Coast from Alaska to Baja*. Palo Alto, Cal.: American West Publishing Company, 1971.

KINGSBURY, JOHN M. *Seaweeds of Cape Cod and the Islands*. Chatham, Mass.: Chatham Press, 1969.

KOZLOFF, EUGENE N. *Plants and Animals of the Pacific Northwest: An Illustrated Guide to the Natural History of Western Oregon, Washington, and British Columbia*. Seattle, Wash.: University of Washington Press, 1976.

————. *Seashore Life of Puget Sound, the Strait of Georgia, and the San Juan Archipelago*. Seattle, Wash.: University of Washington Press, 1973.

KRINSLEY, D. H., AND I. J. SMALLEY. "Sand." *American Scientist*, 60:286–291 (May–June 1972).

LAUFF, GEORGE H. (ED.). *Estuaries*. Washington, D.C.: American Association for the Advancement of Science, 1967.

LEE, THOMAS F. *The Seaweed Handbook: An Illustrated Guide to Seaweeds from North Carolina to the Arctic*. Boston: The Mariners Press, 1977.

LOWERY, GEORGE H., JR. *Louisiana Birds*. 2nd ed. Baton Rouge, La.: Louisiana State University Press, 1960.

————. *The Mammals of Louisiana and Its Adjacent Waters*. Baton Rouge, La.: Louisiana State University Press, 1974.

MACGINITIE, G. E., AND N. MACGINITIE. *Natural History of Marine Animals*. 2nd ed. New York: McGraw-Hill, 1968.

MARX, ROBERT F. *Shipwrecks of the Western Hemisphere*. New York: World Publishing Company, 1972

MORRIS, PERCY A. *A Field Guide to Shells of the Atlantic and Gulf Coasts and the West Indies*. 3rd ed. Boston: Houghton Mifflin Company, 1973.

————. *A Field Guide to Shells of the Pacific Coast and Hawaii*. Boston: Houghton Mifflin Company, 1952.

MUNZ, PHILIP A. *Shore Wildflowers of California, Oregon and Washington*. Berkeley, Cal.: University of California Press, 1964.

NEHLS, HARRY B. *Familiar Birds of Northwest Shores and Waters*. Portland, Ore.: Portland Audubon Society, 1975.

NUTMAN, P. S., AND BARBARA MOSSE (EDS.). *Symbiotic Associations.* New York: Cambridge University Press, 1963.

PALMER, JOHN D. *Biological Clocks in Marine Organisms.* New York: Wiley-Interscience, 1974.

PETRY, LOREN C., AND MARCIA G. NORMAN. *A Beachcomber's Botany.* Chatham, Mass.: The Chatham Conservation Foundation, 1968.

POMEROY, LAWRENCE R. "The Ocean's Food Web, a Changing Paradigm." *Bio-Science,* 24(9):499–504 (September 1974).

POTTER, JEFFREY. *Disaster by Oil.* New York: The Macmillan Company, 1973.

PURCHON, R. D. *The Biology of the Mollusca.* New York: Pergamon Press, 1977.

RICKETTS, EDWARD F., AND JACK CALVIN. *Between Pacific Tides.* 4th ed., revised by Joel W. Hedgpeth. Stanford, Cal.: Stanford University Press, 1968.

RUDLOE, JACK. *The Erotic Ocean: A Handbook for Beachcombers.* New York: World Publishing Company, 1971.

RUGGIERI, GEORGE D. "Drugs from the Sea." *Science* 194:491–497 (29 October 1976).

SAUNDERS, DAVID F. *An Introduction to Biological Rhythms.* New York: Halsted Press, 1977.

SCAGEL, ROBERT F. *Guide to Common Seaweeds of British Columbia.* Handbook No. 27. Victoria, B.C.: British Columbia Provincial Museum, 1967.

SCHWARTZ, MAURICE L. (ED.). *Barrier Islands.* Stroudsburg, Pa.: Dowden, Hutchinson, and Ross, 1973.

SHERWOOD, ARTHUR W. *Understanding the Chesapeake.* Cambridge, Md.: Tidewater Publishers, 1973.

SIMON, ANNE W. *The Thin Edge: Coast and Man in Crisis.* New York: Harper & Row, Publishers, 1978.

SMITH, RALPH I., AND JAMES T. CARLTON (EDS.). *Light's Manual: Intertidal Invertebrates of the Central California Coast.* 3rd ed. Berkeley, Cal.: University of California Press, 1975.

SQUIRE, JAMES L., JR., AND SUSAN E. SMITH. *Anglers' Guide to the United States Pacific Coast.* Seattle, Wash.: Department of Commerce, 1977.

STEPHENSON, T. A., AND ANNE STEPHENSON. *Life Between Tidemarks on Rocky Shores.* San Francisco: W. H. Freeman & Company, Publishers, 1972.

THOREAU, HENRY DAVID. *Cape Cod.* Introduction by Henry Beston. New York: Bramhall House, 1951.

TOWNSEND, CHARLES WENDELL. *Sand Dunes and Salt Marshes.* Boston: The Page Company, Publishers, 1913.

URSIN, MICHAEL J. *Life in and Around the Salt Marshes.* New York: Thomas Y. Crowell Company, 1972.

VERMEER, KEES. *The Breeding Ecology of the Glaucous-winged Gull* (Larus glaucescens) *on Mandarte Island, B.C.* Occasional papers of the British Columbia Provincial Museum. No. 13, 1963.

VOSS, GILBERT L. *Seashore Life of Florida and the Caribbean.* Miami, Fla.: Seemann Publishing, 1977.

WARNER, WILLIAM W. *Beautiful Swimmers: Watermen, Crabs and the Chesapeake Bay.* New York: Penguin Books, 1976.

WATERS, JOHN F. *Exploring New England Shores: A Beachcomber's Handbook.* Lexington, Mass.: Stone Wall Press, 1974.

WOOD, AMOS L. *Beachcombing for Japanese Glass Floats.* Portland, Ore.: Binfords & Mort, Publishers, 1971.

Index

(Page numbers in italics refer to illustrations)